Electronic Circuits for the Evil Genius

Evil Genius Series

123 Robotics Experiments for the Evil Genius

Electronic Gadgets for the Evil Genius: 28 Build-it-Yourself Projects

Electronic Circuits for the Evil Genius: 57 Lessons with Projects

Coming in 2005

123 PIC Microcontroller Experiments for the Evil Genius

Mechatronics for the Evil Genius: 25 Build-it-Yourself Projects

50 Awesome Automotive Projects for the Evil Genius

Solar Energy Projects for the Evil Genius: 16 Build-it-Yourself Thermoelectric and Mechanical Projects

Bionics for the Evil Genius: 25 Build-it-Yourself Projects

MORE Electronic Gadgets for the Evil Genius: 28 MORE Build-it-Yourself Projects

Electronic Circuits for the Evil Genius

DAVE CUTCHER

McGraw-Hill

New York   Chicago   San Francisco   Lisbon   London   Madrid
Mexico City   Milan   New Delhi   San Juan   Seoul
Singapore   Sydney   Toronto

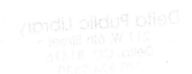

Cataloging-in-Publication Data is on file with the Library of Congress

9 10 11   QPD/QPD   0 1 0 9 8

ISBN-13: 978- 0-07-144881-9
ISBN-10: 0-07-144881-0

*The sponsoring editor for this book was Judy Bass and the production super isor
was Pamela A. Pelton. It was set in Times Ten by MacAllister Publishing Ser ices, LLC.
The art director for the co er was Anthony Landi.*

Printed and bound by Quebecor/Dubuque.

McGraw-Hill books are available at special quantity discounts to use as premiums and sales promotions, or for use in corporate training programs. For more information, please write to the Director of Special Sales, McGraw-Hill Professional, Two Penn Plaza, New York, NY 10121-2298. Or contact your local bookstore.

This book is dedicated to my wife Mary. Over the past two years, even more than before, she has supported, protected, understood, and challenged me. With her many talents, she has confronted me and danced with me. She has inspired me and talked me back to reality. She has directed me, believed in what I was, and nurtured what I could be. Beyond that, she gave me permission and helped me focus. I've seen so many books dedicated to wives, and I never understood why. I thought that maybe such dedications were expected. Now I know. It's one thing to be married to a sports nut and be a sports widow for a season. It's completely different when a technical writer disappears into his own world for months at a time.

About the Author

Dave Cutcher teaches electronics, technology, and industrial arts in Burnaby, British Columbia, Canada. An enthusiastic electronics hobbyist and member of the Vancouver Robotics Club and the Seattle Robotics Society, he shared the plans in this book with fellow teachers. They urged him to write this book.

Contents

Foreword

My name is David De Pieri. I'm a shop teacher and CEO of theShopTeacher.com, a resource center for middle school and high school shop teachers. Previously, I worked as journeyman machinist and CNC programmer/operator for eleven years. As my years in the trade accumulated, I realized I had a strong desire to pass this wealth of knowledge and experience to younger people, so I followed my dream to become a high school shop teacher and subsequently develop my Web site. It was during my schooling in 1990 that I had the great fortune of meeting another equally passionate educator, Dave Cutcher, author of *Electronic Circuits for the Evil Genius*. He was my partner on several projects while at the British Columbia Institute of Technology and his strength, obvious even then, was his ability to explain things clearly.

After teaching machine shop and drafting in my hometown for 12 years and feeling very comfortable, I recently moved to a new community, which meant a new teaching assignment for me with a very new challenge, the electronics program. This was an area where I certainly had questions. I called on Dave, who I hadn't talked to in many years–what a tremendous help he was.

Shop teachers like to share their material and so too, without hesitation, did Dave. He gave me an armful of resource material, including a CD that would make up the backbone of his electronics book. In essence, I had the first copy of his book in digital format. His material was a lifesaver for me. I could see the many hours of hard work he had put into what would become *Electronic Circuits for the Evil Genius*. What I appreciate most about Dave's material is the fact that it works well for the independent learner, like myself, as well as for the kids in the classroom.

Generally speaking, I put my money on the man with experience. Dave Cutcher has been teaching electronics for many years and his format stands out as the only effective introductory electronics books I've seen. His learning curve is gentle but continually challenges the student. One idea builds on the next. Both analog and digital electronics are explained with many hands on and practical projects. His images help explain clearly what words can't show. My students find his work easy to understand and pleasantly taxing.

Over the past few years working with my Web site, project submissions in electronics have been scarce and inquiries have been many as to where someone can get any practical and realistic help. Last year, someone posted a message on my Web site asking if anyone knew of a good text to introduce electronics to the basic beginner. The responses from the members of the Web site were, "Save your money," but Dave responded by saying his book was in the works. With his book completed, I can vouch for him by agreeing this is the only book I've seen that effectively introduces "real" electronics. Using Dave Cutcher's book will eliminate a huge amount of work and frustration that the independent learner has faced until now, and will give them a solid base for understanding electronics. Like I've said on my Web site, "Why do we continue to reinvent the wheel?" We should use what is already established by other teachers. This book greatly lessens the mystery generally associated with electronics. Dave once told me, "Electronics isn't hard . . . it's new, but it isn't hard." He's right, and he proves it with this book.

Electronic Circuits for the Evil Genius provides 57 lessons with good, solid material that makes electronics enjoyable and unintimidating. I highly recommend it to anyone interested in learning electronics at home, or to the classroom teacher as a class set.

David De Pieri
CEO, theShopTeacher.com

Preface

We casually accept electronics in our everyday world. Those who don't understand how it works are casually obedient. Those who take the time to learn electronics are viewed as geniuses. Do you want to learn how to control the power of electronics?

This text provides a solid introduction to the field of electronics, both analog and digital. *Electronic Circuits for the Evil Genius* is based on practical projects that exercise the genius that exists in all of us. Components are introduced as you build working circuits. These circuits are modified and analyzed to help explain the function of the components. It's all hands-on. Analysis is done by observation, using a digital multimeter, and using your computer as an oscilloscope.

You will build two major projects in the first unit:

- An automatic night light
- A professional-quality alarm

The remainder of the text focuses on three major projects, one per unit:

- Building a digital toy using logic gates
- Designing and building an application using digital counting circuits
- Applying transistors and op amps as you build a two-way intercom system

The lessons and prototype circuits built in the book are focused on developing a solid foundation centered on each of these major projects. You work from ideas to prototypes, producing a final product.

I hope you enjoy building the projects and reading the book as much as I enjoyed developing them.

Dave Cutcher

Acknowledgments

For a variety of reasons, there are many people I need to thank.

First are my current guinea pigs, who chose to be caged in a classroom with me for three years running. Andrew Fuller who put together the game "When Resistors Go Bad," which can be found at www.books.mcgraw-hill.com/authors/cutcher. He and André W., two very original evil geniuses. I hope they understand the molar concept in chemistry now and won't raise a stink about me mentioning them. Eric R. and Eric P., both for being the gentler geniuses they are. And Brennen W., who was more patient with me at times than I was with him. It was a difficult year.

I've had only one formal class in electronics, taught by Gus Fraser. He let me teach myself. Bryan Onstad gave me a goal to work toward and platform to work on. Don Nordheimer was the first adult who actually worked through my material outside of the classroom environment. At the same time, he proofed the material from the adult perspective. I owe a heartfelt thanks for the encouragement from Pete Kosonan, the first administrator who enjoyed the creative flow of the students as much as I did. For Steve Bailey, the second administrator I found who wasn't threatened by kids who knew more than he did. For the many others like Paul Wytenbrok, Ian Mattie, Judy Doll, and Don Cann, who continually encouraged me over the five years it took to develop this material. For Brad Thode, who introduced me to the necessity of changing careers within teaching back in 1989. For Mrs. Schluter and Mrs. Gerard, who taught me to believe in myself and recognize that there was room for creativity, not just what they wanted to hear.

Then to Stan Mah. He never explains completely. He sits there with a knowing smile and challenges me. "Think about it before you answer. You can do it," he says. "If I can do it, you can do it."

To my parents, who knew they couldn't change me, so they encouraged me.

Components

In Lesson 1, you will be introduced to many common components that are always present in electronics and many of the bits and pieces you will use in the course. It starts out as a jumble. As you use the parts, the confused mass becomes an organized pile.

In Lesson 2, you become acquainted with the two major tools that you will use throughout the course.

In Lesson 3, you will build your first circuit on the solderless breadboard, a platform that allows you to build circuits in a temporary format.

You use your digital multimeter and get voltage measurements when you set up and test your first circuits.

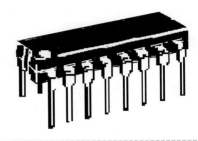

Figure L1-02

Do not remove the small integrated-circuit (IC) chips shown in Figure L1-02 from their antistatic packaging. They are packed in a special antistatic tube or special sponge material.

Lesson 1: Inventory of Parts Used in Part I

All components look the same if you don't know what they are. It's like when you first visit a different country. There's a pile of change, just like in Figure L1-01. You have to be introduced to the currency and practice using it, but you become comfortable with it quickly. Now you need to unjumble the pile and become familiar with your electronic components.

Semiconductors

These are the electronic components you will be using in Part I. As you identify them, set them aside into small groups.

Diodes

You will need three power diodes as shown in Figures L1-03 and L1-04.

The number on the side reads 1N4005. If the last number is not 5, don't worry. Any diode of this series will do the job.

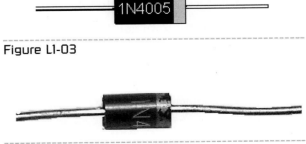

Figure L1-03

Figure L1-04

Figure L1-01

Light-Emitting Diodes (LEDs)

Light-emitting diodes are also known as LEDs. You will need three LEDs. An example is illustrated in Figure L1-05.

Figure L1-05

They can be any color. The most common colors are red, yellow, or green. The color is unimportant.

Resistors

There should be lots of colorful resistors, nearly all the same size.

Notice that in Figure L1-06 each resistor has four color bands to identify it. If you know the colors of the rainbow, you know how to read resistors.

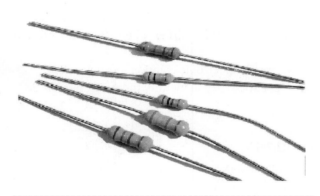

Figure L1-06

Find these resistors:

1 brown-black-brown-gold 100 Ω

2 yellow-violet-brown-gold 470 Ω

1 brown-black-red-gold 1,000 Ω

1 brown-black-orange-gold 10,000 Ω

1 red-red-orange-gold 22,000 Ω

1 yellow-violet-orange-gold 47,000 Ω

1 brown-black-yellow-gold 100,000 Ω

Capacitors

As you notice in Figure L1-07, the capacitor shown is black and white. The colors of capacitors are different depending on the manufacturer. Then again, all pop cans look alike, but each brand has a different label. Locate four small cans, different in size. Written on each are different values and other mumbo jumbo. Look for the information that specifically lists 1 μF, 10 μF, 100 μF, and 1000 μF.

Figure L1-07

There is another capacitor of a different shape to locate. Figure L1-08 shows the other capacitor used in Part I. Again, it is presented in black and white, because the color will change as the manufacturer changes. It is a 0.1 μF capacitor. It may be marked as any of the following: 0.1 or γ1 or 100 nF.

Figure L1-08

Silicon-Controlled Rectifier (SCR)

The ID number 1607B for the *silicon controlled rectifier* (SCR) is written on the face, as shown in Figure L1-09. This SCR comes in this particular package. Not everything with this shape is an SCR, just as not everything in the shape of a pop can is your favorite flavor.

Figure L1-09

Transistors

You need two transistors like that illustrated in Figure L1-10. They are identical except for the numbers 3904 or 3906. All other writing and marks are the manufacturer telling us how great he or she is.

Figure L1-10

Hardware

The solderless breadboard is shown in Figure L1-11.

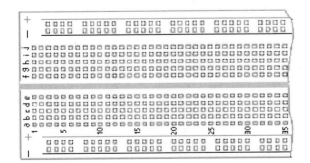

Figure L1-11

Figures L1-12 and L1-13 illustrate two push buttons—they are different, but you can't tell by looking at them. Figure L1-12 is the normally open push button (push to close the contacts), and Figure L1-13 shows the normally closed push button (push to open the contacts).

Figure L1-12

Figure L1-13

You should have lots of 24-gauge solid wire with plastic insulation. There should be plenty of different lengths.

Two battery clips are shown in Figure L1-14.

Figure L1-14

A 9-volt buzzer is shown in Figure L1-15.

Figure L1-15

Two printed circuit boards are premade for your projects: Figure L1-16 is to be used for the night-light project; Figure L1-17 is for your SCR alarm project.

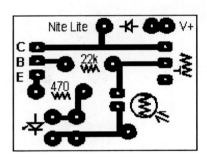

Figure L1-16

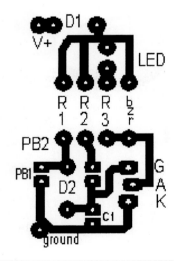

Figure L1-17

Two adjustable resistors are also supplied: The *light-dependent resistor* (LDR) is shown in Figure L1-18 and the potentiometer is shown here in Figure L1-19.

Figure L1-18

Figure L1-19

Lesson 2: Major Equipment

The Solderless Breadboard

When smart people come up with ideas, first they test those ideas. They build a prototype. The easiest way to build prototypes and play with ideas in electronics is on the solderless breadboard, shown here in Figure L2-01.

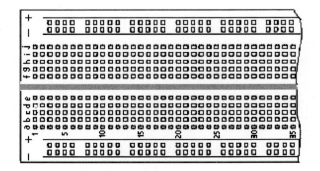

Figure L2-01

Lesson 1 – Inventory of Parts Used

The main advantage of the solderless breadboard is the ability to exchange parts easily and quickly.

The top view in Figure L2-01 shows the many pairs of short five-hole rows and a pair of long rows down each side; each of these lines are marked with a strip of paint.

The Digital Multimeter

I recommend the Circuit Test DMR2900 displayed in Figure L2-02.

Figure L2-03

Figure L2-02

The auto-ranging DMM offers beginners the advantage of being easier to learn.

The second style of DMM is not auto-ranging. This style is easy to use after you become familiar with electronics, but they tend to be confusing for the beginner. A typical dial of a nonautoranging multimeter is confusing, as you can see in Figure L2-03.

I discourage the use of outdated whisker-style multimeters for this course. Figure L2-04 gives an example of what to avoid.

Figure L2-04

Connection Wire

A box of wire provided in the kit is displayed in Figure L2-05.

Figure L2-05

These are different lengths convenient for the solderless breadboard. If you need to cut the wire to different lengths, wire clippers will work perfectly. Old scissors work as well.

Set the dial of the DMM to CONTINUITY. This setting is shown in Figure L2-06.

Figure L2-06

Touch the end of both red and black probes to the colored covering. The DMM should be silent and read OL, as in the readout illustrated in Figure L2-07, because the resistance of the insulation prevents any current from passing.

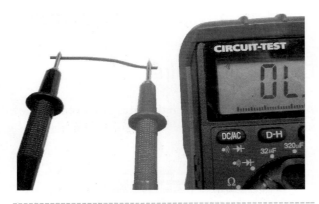

Figure L2-07

Be sure the strip of insulating plastic is removed from both ends of the piece of wire as demonstrated in Figure L2-08. If you don't have a proper wire stripper available, use a knife or your fingernails to cut the insulation. Be careful not to nick the wire inside the insulation.

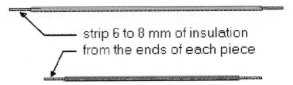

strip 6 to 8 mm of insulation
from the ends of each piece

Figure L2-08

Now touch the end of both probes to the exposed wire. The DMM should read "00" and beep, just like the readout in Figure L2-09. The wire is a good conductor, and the DMM shows "continuity," a connected path.

Figure L2-09

Exercise: Mapping the Solderless Breadboard

Strip the end of two pieces of wire far enough to wrap around the DMM probes on one end and enough to insert into the SBB on the other end, as shown in Figure L2-10.

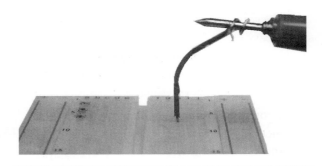

Figure L2-10

1. Set your digital multimeter to continuity. Now refer to Figure L2-11. Notice the letters across the top and the numbers down the side of the solderless breadboard.

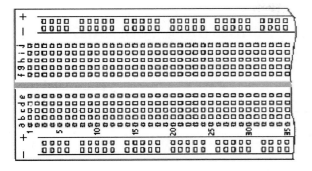

Figure L2-11

2. Probe placement
    a. Place the end of one probe wire into the SBB at point "h3" and mark that on the drawing.
    b. Use the other probe to find three holes connected to the first. The multimeter will indicate the connection.
    c. Draw these connections as solid lines.

3. Base points
    a. Create four more base points at e25, b16, f30, and c8.
    b. Use the other probe to find three holes connected to each of these points.
    c. Again draw these connections as solid lines.

4. Additional base points
    a. Choose two more base points on the outside long, paired lines. These lines are not lettered or numbered but have a stripe of paint along the side. Mark them on the diagram above.
    b. Find three holes connected to each of these points.
    c. Again draw these connections as solid lines.

5. Be sure that you can define the terms prototype, insulator, and conductor.

6. With your multimeter set on continuity, walk around and identify at least five common items that are insulators and five common materials that are conductors.

Lesson 3: Your First Circuit

As you see, the solderless breadboard has a definite layout as shown in Figure L3-01. One strip of the spring metal in the breadboard connects the five holes. You can easily connect five pieces in one strip. The two long rows of holes allow power access along the entire length of the breadboard.

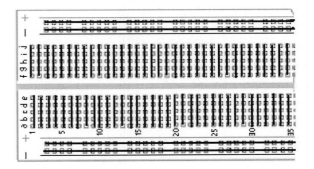

Figure L3-01

Setting Up the Solderless Breadboard

You will have a standard set up for every circuit. The battery clip is connected to one of the first rows of the breadboard, and the diode connects that row to the outer red line. See Figure L3-02.

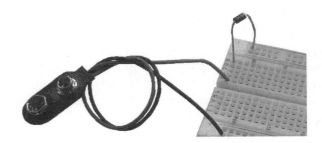

Figure L3-02

Notice the gray band highlighted in Figure L3-03 on the diode. It faces in the direction that the voltage is pushing.

Figure L3-03

The voltage comes through the red wire, through the diode, and then to the power strip on the breadboard.

Why Bother?

This power diode provides protection for each circuit that you build in the following ways:.

- The diode is a one-way street. You can view the animated version of Figure L3-04 at the Web site www.books.mcgraw-hill.com/authors/cutcher.

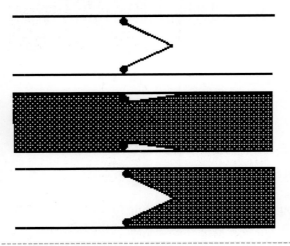

Figure L3-04

- Many electronic components can be damaged or destroyed if the current is pushed through them the wrong way, even for a fraction of a second.
- This standard breadboard setup helps ensure you will always have your battery connected properly.

- If you accidentally touch the battery to the clip backwards, nothing will happen, because the diode will prevent the current from moving.

Breadboarding Your First Circuit

Parts List

D1—Power diode 1n400x

LED1—LED any color

R1—470-ohm resistor

Your LED is a light-emitting diode. That's right, a diode that emits light. It has the same symbol as a diode, but it has a "ray" coming out, as shown here in Figure L3-05.

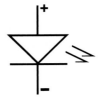

Figure L3-05

Figure L3-06 is a picture of an LED. Never touch your LED directly to your power supply. A burned-out LED looks just like a working LED. Note in the picture how to identify the negative side.

Figure L3-06

- The shorter leg: This is always reliable with new LEDs, but not with ones that you have handled in and out of your breadboard. As you handle the components, the legs can get bent out of shape.

- The flat side on the rim: This is always reliable with round LEDs, but you have to look for it.

Remember, that the LED, as a diode, is a one-way street. It will not work if you put it in backward.

Figure L3-07 shows several resistors. The resistor symbol is illustrated in Figure L3-08. The resistor you need is the 470-ohm yellow-violet-brown-gold.

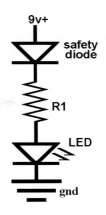

Figure L3-09

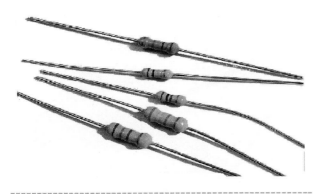

Figure L3-07

Figure L3-08

Resistance is measured in ohms. The symbol for ohms is the Greek capital letter omega, Ω.

The schematic is shown in Figure L3-09. Set up your breadboard as shown in Figure L3-10. Note that this picture shows the correct connections. The red wire of the battery clip is connected to the power diode that in turn provides voltage to the top of the breadboard. The black wire is connected to the blue line at the bottom of the breadboard.

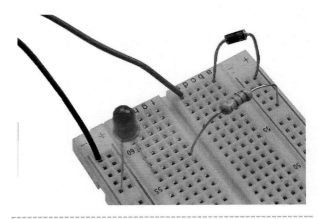

Figure L3-10

A Quick Note

1. Always complete your breadboard before you attach your power to the circuit.

2. Attach your battery only when you are ready to test the circuit.

3. When you have finished testing your circuit, take your battery off.

When you think you've got it, connect the battery and find out.

Your First Circuit Should Be Working

Figure L3-11 shows what is happening. Like a waterfall, all of the water goes from the top to the bottom. The resistor and LED each use up part of the voltage. Together they use all the voltage. The 470-ohm resistor uses enough voltage to make sure the LED has enough to work, but not so much that would burn it out.

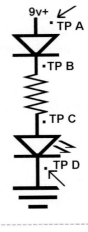

Figure L3-12

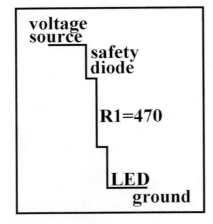

Figure L3-11

Let's look at how the voltage is being used in the circuit

1. Set the DMM to *direct current voltage* (DCV). If you are using a multimeter that is not auto-ranging, set it to the 10-volt range.

2. Measure the voltage of the 9-volt battery while it is connected to the circuit.

3. Place the red (+) probe at test point A (TP-A) and the black (−) probe at TP-D (ground). The arrows in the schematic shown in Figure L3-12 indicate where to attach the probes. Corresponding test points have been noted in Figure L3-13 as well.

4. Record your working battery voltage.   ____V

5. Measure the voltage used between the following points:

   TP-A to TP-B across the safety diode   ____V

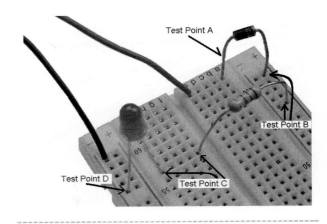

Figure L3-13

   TP-B to TP-C across the 470-ohm resistor   ____V

   TP-C to TP-D across the LED   ____V

6. Now add the voltages from #5.   ____V

7. List working battery voltage (recorded in item 2).   ____V

8. Compare the voltage used by all of the parts to the voltage provided by the battery.

The voltages added together should be approximately the same as the voltage provided by the battery. It may be only a few hundredths of a volt difference.

Section Two

Resist If You Must

Lesson 4: Reading Resistors

If you know the colors of the rainbow, you know how
to read resistors (Table L4-1).

Brown   **Red**   **Orange**   **Yellow**   **Green**   **Blue**   **Violet**   **Gray**   **White**

The gold bands are always read last. They indicate
that the resistor's value is accurate to within 5%.

Table L4-1   Resistor band designations

Color Band	First Band: Value	Second Band: Value	Third Band: Number of Zeros	Units
Black	0	0	No zeros	Tens ##
Brown	1	1	One zero "0"	Hundreds ##0
Red	2	2	Two zeros "00"	Thousands (k) #, #00
Orange	3	3	Three zeros "000"	Ten thousands (k) ##,000
Yellow	4	4	Four zeros "0,000"	Hundred thousands (k) ##0,000
Green	5	5	Five zeros "00,000"	Millions (M) #,#00,000
Blue	6	6	Six zeros "000,000"	Ten millions (M) ##,000,000
Violet	7	7	Not available	
Gray	8	8	Not available	
White	9	9	Not available	

13

When using the digital multimeter to measure resistance, set the dial to Ω. Notice the two points of detail shown in Figure L4-1.

Figure L4-1

The first point is that when the dial is set directly to the Ω symbol to measure resistance, it also appears on the readout. Secondly, notice the M next to the Ω symbol. That means the resistor being measured is 0.463 MΩ. That is 0.463 million ohms, or 463,000 ohms. When it is there, *never* ignore that extra letter.

As you use resistors, you quickly become familiar with them. The third band is the most important marker. It tells you the range in a power of 10. In a pinch, you could substitute any resistor of nearly the same value. For example, a substitution of a red-red-orange could be made for a brown-black-orange resistor. But a substitution of a red-red-orange by a red-red-yellow would create more problems than it would solve. Using a completely wrong value of resistor can mess things up.

Exercise: Reading Resistors   If you have an auto-ranging multimeter, set the DMM to measure Resistance. If you do not have an auto ranging DMM, you have to work harder because the resistors come in different ranges. You have to set the range on your DMM to match the range of the resistor. That means that you should have an idea of how to read resistor values before you can measure those values using a DMM that is not auto ranging. An auto ranging DMM really does make it much easier.

Your skin will conduct electricity, and if you have contact with both sides of the resistor, the DMM will be measuring your resistance mixed with the resistor's. This will give an inaccurate value.

Proper Method to Measure Resistor's Value   Figure L4-2 shows how to measure a resistor. Place one end of the resistor into your solderless breadboard and hold the probe tightly against it, but not touching the metal. You can press the other probe against the top of the resistor with your other finger.

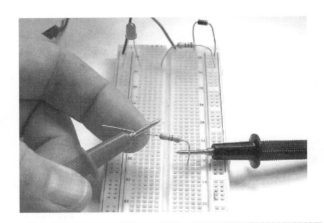

Figure L4-2

1. Some of the resistors you will need to be able to identify, because you use them soon, are listed in Table L4-2.

Table L4-2 Resistors needed

First Band: Value	Second Band: Value	Third Band: Number of Zeros	Resistor Value	DMM Value
Brown 1	Black 0	Brown 0	100 Ω	_____ Ω
4	Violet 7	Brown 0	470 Ω	_____ Ω
Brown ____	Black ____	Red 00	1,000 Ω	_____ Ω
Brown ____	Black ____	Orange 000	10 kΩ 10,000Ω	_____ Ω
Red ____	Red ____	Orange ____	22 kΩ 22,000 Ω	_____ Ω
Brown ____	Black ____	Yellow ____	100 kΩ 100,000	_____ Ω

2. Don't be surprised if the resistor value is not exactly right. These resistors have a maximum error of 5%. That means that the 100-ohm resistor can be as much as 105 ohms or as little as 95 ohms. Plus or minus 5 ohms isn't too bad. What is 5% of 1,000,000?

   What is the maximum you would expect to see on the 1,000-ohm resistor? _____ Ω

   What is the minimum you would expect to see on the same 1-kilo-ohm resistor? _____ Ω

3. Measure your skin's resistance by holding a probe in each hand. It will bounce around, but try to take an average. _____ Ω

   Did you know that this can be used as a crude lie detector? A person sweats when they get anxious. Have a friend hold the probes. Then ask them an embarrassing question. Watch the resistance go down for a moment.

4. Write each of these values as a number with no abbreviations.

   10 kΩ = _____ Ω

   1 kΩ = _____ Ω

   .47 kΩ = _____ Ω

   47 kΩ = _____ Ω

Lesson 5: The Effect Resistors Have on a Circuit

Let's go back to the breadboard and see how different resistors affect a simple circuit. The resistors and the LEDs are both loads. The resistor uses most of the voltage, leaving just enough for the LED to work. The LEDs need about 2 volts.

What would happen if you changed resistors on the circuit you just built, shown in Figure L5-1?

Figure L5-1

You measured the voltage used across the resistor from TP-B to TP-C and measured the voltage used across the LED from TP-C to TP-D.

Figure L5-2 is the schematic of the circuit.

Figure L5-3 shows a waterfall. A waterfall analogy explains how voltage is used up in this circuit. The water falls over the edge. Some of the force is used up by the first load, the safety diode. More of the voltage is then used by the second load, the resistor. The remaining voltage is used by the LED.

This "waterfall" shows how the voltage is used by a 470-ohm resistor. If the resistor wasn't there, the LED would be hit with the electrical pressure of more than 8 volts. It would burn out. The waterfall analogy helps you visualize how voltage is used in a circuit.

Remember, all the water over the top goes to the bottom, and all of the voltage is used between source and ground. Each ledge uses some of the force of the falling water. Each component uses part of the voltage.

What happens if there is more resistance? More of the voltage is used to push the current through that part of the circuit, leaving less to power the LED.

This is represented visually in Figure L5-4.

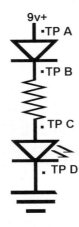

Figure L5-2

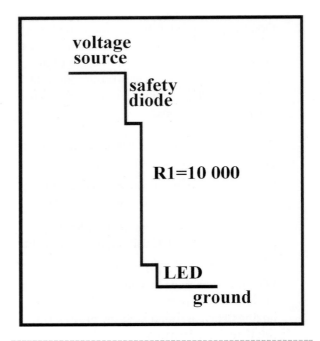

Figure L5-4

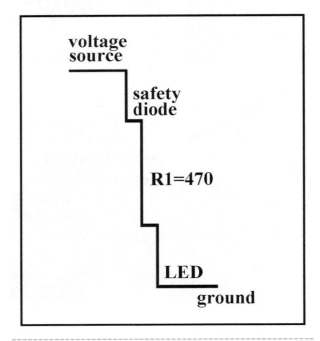

Figure L5-3

Exercise: The Effect Resistors Have on a Circuit

Your setup should look like Figure L5-5. Have your resistors arranged from lowest to highest value as presented in Table L5-1.

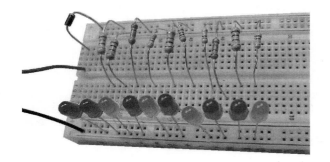

Figure L5-5

Table L5-1   Exercise sheet

Resistor Value	Total Voltage Available	Voltage Drop Across Resistor	Voltage Drop Across the LED	LED Brightness (compared to 470 Ω)
100 Ω	____V	____V	____V	_____
470 Ω	____V	____V	____V	**Normal**
2,200 Ω	____V	____V	____V	_____
10,000 Ω	____V	____V	____V	_____
47,000 Ω	____V	____V	____V	_____
220,000 Ω	____V	____V	____V	_____

Lesson 6: The Potentiometer

Not all resistors are "fixed" like the small color-banded ones that you've already been introduced to. A common variable resistor is the potentiometer, pictured in Figure L6-1.

This useful device is often simply referred to as a pot. A smaller version is also shown. These are called *trim pots*. You have often used potentiometers as volume controls. The maximum resistance value is usually stamped onto the metal case.

Figure L6-1

Figure L6-2 shows a picture of a potentiometer taken apart. The potentiometer works because the sweep arm moves across the carbon ring and connects that to the center. The leg on the left is referred to as A, the center leg as C (center), and the right leg as B.

Figure L6-2

The carbon ring shown in Figure L6-3 is the heart of the potentiometer. It is made of carbon mixed with clay. Clay is an insulator. Carbon is the conductor.

The action of the potentiometer is the sweep arm (copper on white plastic) moving across the carbon ring (Figure L6-4). The sweep arm allows the current to move between A and C as its position changes. The resistance between A and C also changes with distance.

Figure L6-3

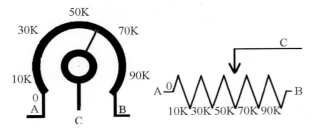

Figure L6-4

The distance between A and B is always the same, so the resistance between A and B is always the same. The value for this demonstration potentiometer is 100,000 ohm. The 100-kilo-ohm value means the set value between legs A and B is 100 kilo-ohm. Ideally, the minimum between A and C is 0 ohm (directly connected), and the maximum between A and C should be 100 kilo-ohm.

The ratio between carbon and clay determines how easily electrons pass through the a resistor. More clay mixed in leaves less carbon. The less carbon means less conducting material. That creates higher resistance.

The carbon in the ring is similar to the carbon in a pencil. The pencil lead is also made of a mixture of carbon and clay. Carbon is the conductor. Clay is the insulator.

Soft pencils have less clay, and more carbon. A mark by a soft pencil will have less resistance.

Hard pencils have lead that contains more clay and less carbon. These provide higher resistance.

1. Use a No. 2 soft pencil to draw a thick line on this piece of paper as demonstrated in Figure L6-5. A harder pencil has too much clay and will not give good results.

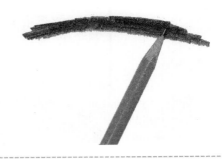

Figure L6-5

2. Set your multimeter to measure resistance Ω. If it is not auto ranging, set it to maximum resistance.

3. Just as it is shown in Figure L6-6, press the probes down hard against the pencil trace about an inch apart. Be sure that you don't touch the tips of the probe. You want to measure the resistance of the pencil trace, not the resistance of your body.

Figure L6-6

a. Now record the resistance from the multimeter. _____ Ω If the DMM says the resistance is out of range, move the probes together until you get a reading.

b. Move the probes closer together; then further apart. Write down what you observe.

Lesson 6 – The Potentiometer

4. Use the 100-kilo-ohm potentiometer. Record your results.

   a. Measure the resistance between the two outer legs A and B. _____ Ω

   b. Adjust the knob and check the resistance between A and B again. _____ Ω

   c. Adjust the knob about 1/2 way. Measure the resistance between the left and middle leg—A and C. _____ Ω

   d. Turn the knob a bit and check again. Note any change. _____ Ω

   Explain what is happening, relating that to the carbon ring shown in Figure L6-3.

   _____

   _____

Breadboarding the Circuit

Note the similarities of the schematic shown in Figure L6-7 and the picture of the circuit displayed in Figure L6-8.

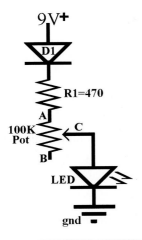

Figure L6-7

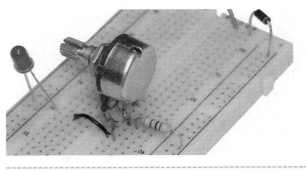

Figure L6-8

5. Make sure that you have the battery hooked up properly through the power diode placed properly as noted on the schematic.

6. As you turn the shaft of the potentiometer, the LED should brighten and dim. Explain what is happening.

   _____

   _____

7. Why is there a 470-ohm fixed resistor in this circuit?_____

Lesson 7: Light-Dependent Resistors

Another variable resistor is the *light-dependent resistor* (LDR). The LDR changes its ability to conduct electrons with the change of light. It is commonly used to turn equipment on automatically as night falls. Some cars use it as the input to the switch that turns on headlights as conditions change, even as they drive through a tunnel. The symbol for the LDR is shown here in Figure L7-1.

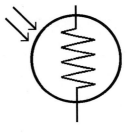

Figure L7-1

There is no room to place a value on most LDRs. They are ordered and supplied in specific values. An easy way to measure the maximum resistance is to measure it in darkness.

Insert the LDR onto the breadboard so the legs are not connected, as shown here in Figure L7-2. Measure the resistance using your DMM. The readout may be jumping around because LDRs are sensitive.

Figure L7-2

Look at Figure L7-3. Place lid of a black pen over the LDR and measure the resistance again.

Figure L7-3

Breadboard the Circuit in Table L7-1

Table L7-1 Example circuit

Parts List

D1	Power diode
LDR	1 MΩ dark
LED	5 mm round

Note the similarities of the schematic in Figure L7-4 and the breadboard layout in Figure L7-5.

Figure L7-4

Figure L7-5

What to Expect

1. Attach the battery and note the brightness of the LED. It should be fairly bright.

2. Place the lid of the pen over the LDR again. The LED should dim to nearly nothing.

3. Consider this. What is the relationship between the amount of light on the LDR and the LDR's resistance?

Exercise: Light-Dependent Resistors

1. Disconnect the power supply. Measure and record the resistance of the LDR in the light. It may be necessary to take a rough average because it will be jumping around wildly.

2. Place a dark black pen lid over the LED and measure the resistance again. Remember that your fingers can affect the read out.

3. Attach the power supply and note the brightness of the LED. Place the lid of the pen over the LDR again. State the relationship between the amount of light on the LDR and the resistance of the LDR.

4. Note the minimum resistance that occurs on the LDR in the light. Why is the 470-ohm resistor not used in this circuit?

5. Consider the "waterfall" diagrams presented in Figure L7-6. From brightest to darkest conditions, what would be the best order of these diagrams regarding the LDR's effect on the brightness of the LED?

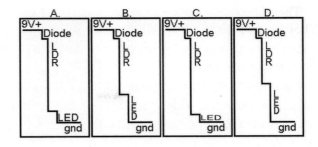

Figure L7-6

Figure L8-2

More Components and Semiconductors

Lesson 8: Capacitors and Push Buttons

Yes, there is more to electronics than resistors and LEDs. Capacitors are used to store small charges. Push buttons allow you to control connections to voltage. This lesson introduces both capacitors and push buttons. You then build a circuit that applies them together.

Capacitors

A capacitor has the capacity (ability) to store an electric charge. You can see in Figure L8-1 that the symbol of the capacitor represents two plates.

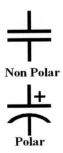

Non Polar

Polar

Figure L8-1

In Figure L8-2, the opened capacitor clearly shows that the capacitor is made of just two metal plates, with a bit of insulation between them. They come in three basic shapes and all sizes.

Capacitors in the upper range, 1 microfarad and higher, are electrolytic capacitors. They must be connected the right direction. There are two indicators of the negative side. First, there is a colored stripe down the side that indicates polarity, and second, if both legs come out of the same side, one leg is shorter. That is the negative leg. It is "minus" some length. Only the electrolytic capacitors have a positive and negative side. The disk and film capacitors do not have a positive or negative side. A variety of capacitors are represented in Figure L8-3.

Figure L8-3

Remember that a backwards electrolytic is a dead electrolytic. These must be connected incorrectly. Figure L8-4 helps remind us.

Figure L8-4

When electricity was first being defined over 200 years ago, the measurements were done with crude instruments that were not sensitive. The people who defined the units missed the mark, but we still use them today. The farad is the basic unit of capacitance. One farad is so huge that today, the standard unit in electronics is one millionth of a farad. The Greek let-

ter μ (mu) represents micro for the unit. That is 0.000001 F or 1×10^{-6} farads and is commonly written as $1\ \mu F$ ($1\ \mu F = 1$ microfarad $= 0.000001$ F $= 1 \times 10^{-6}$ F).

We'll go back to using the water analogy. If you think of the electric charge like water, the capacitors can be compared to containers able to hold that water. Capacitors have the ability (capacity) to store an electric charge.

The amount of charge capacitors can hold depends on their purpose, just like varying size containers used to hold water. Such containers are pictured in Figures L8-5, L8-6, and L8-7.

Figure L8-5

Figure L8-6

Figure L8-7

As mentioned before, capacitors come in three standard types.

Disk capacitors hold the smallest amount. They have a common shape shown in Figure L8-8. They are so small that their capacitance is measured in trillionths of a farad, called *picofarads*. Their general range is from 1 picofarad to 1,000 picofarads. To look at that another way, that is 1 millionth of a microfarad to 1 thousandth of a microfarad.

Figure L8-9

Figure L8-8

To visualize the size of charge they are able to hold, think of water containers ranging from a thimble (Figure L8-9) up to a mug (Figure L8-10)

Figure L8-10

Film capacitors are box shaped as shown in Figure L8-11. They are midrange. They hold between a thousandth of a microfarad up to a full microfarad.

Figure L8-11

Their capacitance range is 1,000 times that of the disk capacitor. A good analogy for the relative size of charge a film capacitor holds is to think of a sink shown in Figure L8-12 up to the size of a large bathtub in Figure L8-13.

Figure L8-12

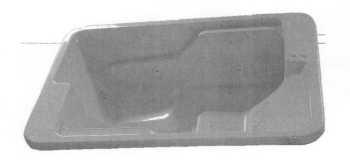

Figure L8-13

Electrolytic capacitors are small and can-shaped. Find the electrolytic capacitors in your inventory. They should look similar to the electrolytic capacitors pictured in Figure L8-14. There might be various colors.

These hold the larger amounts of 1 microfarad and above. Their capacitance abilities can be thought of in larger dimensions from swimming pools (Figure L8-15) to lakes (Figure L8-16).

Figure L8-14

Figure L8-15

Figure L8-16

Push Buttons

There are two main types of push buttons, but they can both look identical to the picture in Figure L8-17.

Figure L8-17

Push Button Normally Open (PBNO)

Push the button; a piece of metal connects with two metal tabs inside as you can see in Figure L8-18. It creates a temporary path and the charge can flow. Set your DMM to continuity and put a probe to each contact for the push button. Continuity should show only when you are pushing the plunger down.

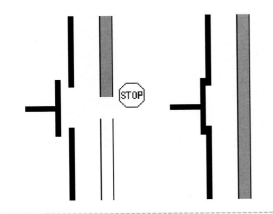

Figure L8-18

Push Button Normally Closed (PBNC)

Push the button; a piece of metal disconnects from the two metal tabs inside as depicted in Figure L8-19. It creates a temporary break and the charge cannot flow. Set your DMM to continuity and put a probe to each contact for the push button. Continuity will show all the time, except when you are pushing the button down.

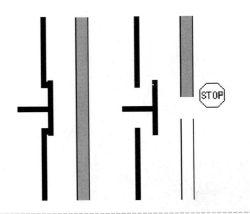

Figure L8-19

Build This Circuit

Build the circuit shown in Figure L8-20 (see also Table L8-1). Note the similarity between the schematic in Figure L8-20 and the photograph in Figure L8-21.

Table L8-1   Parts list for circuit in Figure L8-20

Parts List	
PB1	Normally open Solder connecting wire to the legs
C1	1,000 µF electrolytic
LED	5 mm round
R1	470 Ω

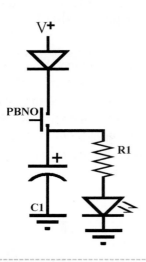

Figure L8-20

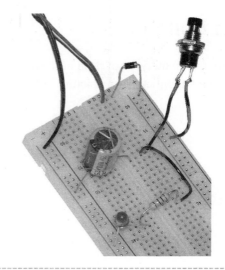

Figure L8-21

How It Works

Carefully note the sequence of actions in Figure L8-22.

1. The normally open push button closes.

2. Voltage fills the capacitor and powers the LED.

3. The PB NO opens, cutting off the voltage.

4. The capacitor drains through the LED.

    a. As the capacitor drains, the voltage decreases.

    b. As the voltage decreases, the LED dims.

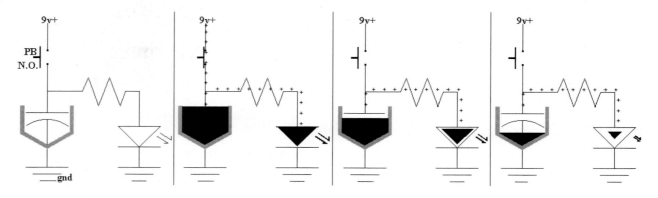

Figure L8-22

Exercise: Capacitors and Push Buttons

1. Look closely at the electrolytic capacitors. Be sure to note the stripe and the short leg that marks the polarity.

2. Describe what happens in your circuit as you push the button, then let go.

3. a. Disconnect the wire indicated in Figure L8-23 between the capacitor and R1.

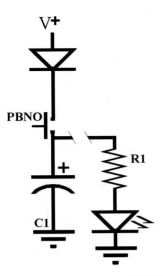

Figure L8-23

b. Push the button to charge the capacitor. Now wait for a minute or so.

c. Set your DMM to the proper voltage range. Put the red probe to the positive side of the cap, and the black probe to ground. Record the voltage that first appears . The capacitor will slowly leak its charge through the DMM.

d. Reconnect the wire and describe what happens.

4. a. Use the Table L8-2 to record your information as you play with your circuit. As you replace each capacitor and record the time, the LED stays on. Don't expect the time to be very exact.

Table L8-2   Information record

Cap Value	TIME
1,000 μF	
470 μF	
100 μF	
10 μF	
1 μF	

b. Describe the pattern that you see here.

_____

5. Briefly describe what capacitors do.

_____

6. Place the 1,000-microfarad capacitor back into its original position. Now replace the normally open push button (PBNO) with the normally closed push button (PBNC). Describe the action of this circuit.

_____

Lesson 9: Introducing Transistors

Remember: Learning electronics is not hard. It is lots of new information, but it is not hard. Think about it, but not as hard as the guy in Figure L9-1.

Figure L9-1

Considering that it has only been a bit more than 100 years since the first transatlantic radio message, electronics is a very young technology. The invention of the transistor in 1947 was the first step towards micro sizing of all electronics we use today. The NPN is truly electronic. It acts like a normally open push button, but has no moving parts. The transistor is the basic electronic switch. It does have an interesting history that makes for good outside reading. Our entire electronic age is dependent on this device.

Transistors are commonly packaged in the TO-92 case shown in Figure L9-2. Notice how the legs correspond to the schematic symbol in Figure L9-3.

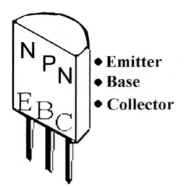

Figure L9-2

Figure L9-3

Note the arrow inside the schematic symbol. It indicates two things: First, it points in the direction of the current, towards ground. Second, it is always on the side of the emitter. It is important to identify the legs of the transistor. For this package, it is easy to remember. Hold the transistor in your fingers with the flat face toward you. Think of a high mountain in the rugged back country of British Columbia—a cliff face reaching skyward. Now, reading left to right, whisper Enjoy British Columbia. You have just identified the three legs. Cute, but it helps.

There are thousands of different types of transistors. The only way to them is to read the numbers printed on the face of the package itself. But even with thousands, there are only two basic types of transistors, the NPN transistor and the PNP transistor.

The NPN Transistor

This lesson introduces the NPN transistor, using the 3904 NPN. Lesson 10 introduces the 3906 PNP. They are opposites but evenly matched in their properties.

Figure L9-4

The NPN transistor is turned on when voltage and current is applied to the base. The NPN transistor acts very much like a water faucet pictured in Figure L9-4. A little pressure on the handle opens the valve, releasing the water under pressure.

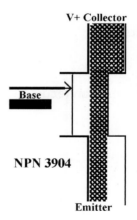

Figure L9-5

Lesson 9 — Introducing Transistors

As you can see in Figure L9-5, a little voltage and current on the base of the NPN transistor leads to a very large increase in the flow of current through the NPN transistor from the collector to the emitter.

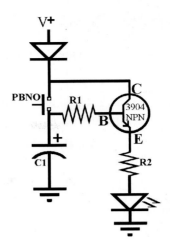

Figure L9-7

Figure L9-6

Another way of thinking about it—the force needed to open the gates on the Grand Coulee Dam, pictured in Figure L9-6, is small compared to the amount of force that moves through that gate.

Build the NPN Transistor Demonstration Circuit

You have used the capacitor to store small amounts of electricity. It powered the LED directly, but could only do that for a brief moment. Here, we use the capacitor to power the transistor. Again, you need to note the similarity between the schematic in Figure L9-7 and the way the circuit is pictured on the solderless breadboard in Figure L9-8 (see also Table L9-1).

Table L9-1   Parts list for Figure L9-8

Parts List	
D1	Safety diode
PB	PBNO
C1	10 μF
R1	22 kΩ
R2	470 Ω
Q1	NPN 3904
LED	5 mm round

Figure L9-8

What to Expect

The LED stays off as you attach your battery. Push and release the push button. The LED will turn on immediately. It will dim and turn off. This action is faster with smaller capacitors.

How This Circuit Works

You are using the charge held in the capacitor to power the transistor. The transistor provides a path for the current to the LED.

Because the base of the transistor uses much less power than the LED, the voltage drains from the capacitor very slowly. The higher value resistor of 22,000 ohms slows the drain from the capacitor significantly.

The LED staaaaaays on muuuuuuch loooooonger.

Exercise: Introducing Transistors

1. Briefly describe the purpose of the transistor.

   _____

2. What do you think? Anything that looks like a transistor is a transistor.

3. Describe how to tell which leg of the transistor is the emitter.

   _____

4. Which leg of the transistor is the base?

   _____

5. What two separate things does the arrow inside the transistor symbol indicate.

   a. _____

   _____

   b. _____

   _____

6. What is the only way to tell the type of transistor? _____

   _____

7. Regarding the water faucet analogy, is the water pressure provided by the water system or the handle? The pressure is provided by the

   _____.

Press and release the push button.

8. After you release the push button, what part provides the power to the base of the transistor? _____

   _____

9. Describe the path of the current that provides the power to the LED. Here is something to consider regarding the answer. The capacitor is not powering the LED. It is only powering the transistor.

10. Record three time trials of how long the LED stays on with the 10-microfarad capacitor.

Time 1	Time 2	Time 3	Average
___ s	___ s	___ s	___ s

Replace C1 with the 100-microfarad capacitor. Time the LED here for three times as well, and find the average.

Time 1	Time 2	Time 3	Average
___ s	___ s	___ s	___ s

Roughly stated, how much more time did the 100-microfarad capacitor keep the LED on than the 10-microfarad capacitor?

a. Three times longer

b. Five times longer

c. Eight times longer

d. Ten times longer

Write down your prediction of how much time the 1,000-microfarad capacitor would *keep* the LED working.

OK, now put in your 1,000-microfarad capacitor. Try it out, three times and average the time.

Time 1	Time 2	Time 3	Average
___ s	___ s	___ s	___ s

How accurate was your prediction?

Describe in detail how this circuit works. Consulting Figure L8-22 of the capacitor powering the LED once the push button is released, the voltage pressure to the base is provided by the

_____.

Lesson 10: The PNP Transistor

This page introduces the PNP transistor, using the 3906 PNP. The previous page introduced the 3904 NPN. They are opposites, but evenly matched in their properties.

The identity of the legs on the TO92 package stay the same as shown in Figure L10-1. But look closely at the symbol for the PNP in Figure L10-02.

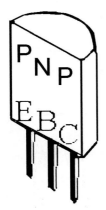

Figure L10-1

Figure L10-2

Note that the schematic symbol of the transistor holds some very important information. The arrow inside the schematic symbol still points in the direction of the common current, but is on the top side now. Because it is always on the side of the emitter, that means that the PNP emitters and collectors have reversed positions relative to the NPN. The legs on the package are still the same, though. The emitter and collector have traded positions relative to the current flow.

Not only are the emitter and collector positions reversed, but the action is reversed as well. The PNP transistor's action is opposite of the NPN. As you increase the voltage to the base, the flow decreases; and as the voltage to the base decreases, the PNP transistor is turned on more. The valve starts in an open position.

The PNP transistor still acts very much like the water faucet shown in Figures L10-3 and L10-4. But now, a little pressure on the handle closes the valve, stopping the water. No pressure on the handle allows the water to push through the faucet. No pressure (voltage) on the base of the PNP transistor allows the voltage and current to push through the transistor.

Figure L10-3

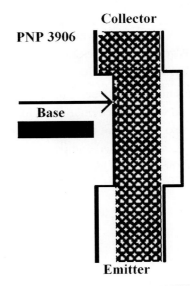

Figure L10-4

But just like turning the water faucet's handle will decrease the water flow, voltage to the base of the transistor will decrease the flow of current through the transistor. Enough pressure on the handle of the water faucet will shut it off. Enough voltage at the base will turn the PNP transistor off completely as well.

Surprisingly, the base has the same action for both the NPN and PNP transistors. It just has a different starting position as shown in Figure L10-5.

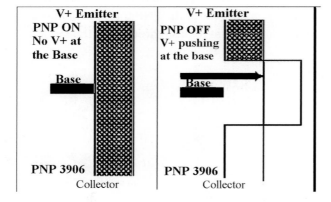

Figure L10-5

Breadboard the PNP Transistor Demonstration Circuit

The capacitor is powering the transistor. But remember for this PNP transistor, when the capacitor is charged, it is going to put pressure on the base of the transistor that will stop the flow.

Note the similarity between this schematic in Figure L10-6 and what the schematic was for the NPN transistor. Also, notice that the transistor in Figure L10-7 is physically reversed compared to the NPN transistor in the previous lesson.

Table L10-1   Parts list for Figure L10-6

Parts List	
Q1	PNP 3906
R1	100 kΩ
R2	22 kΩ
R3	470 Ω
C1	10 μF
PB	N.O.
LED	5 mm round

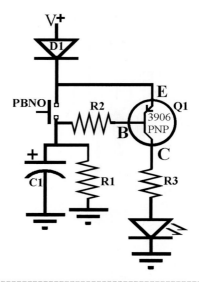

Figure L10-6

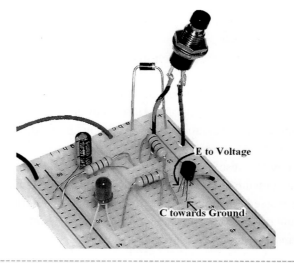

Figure L10-7

What to Expect

The LED turns on as soon as you attach your battery.

Push and release the push button.

The LED will turn off immediately. It will slowly turn back on.

How It Works

1. When you first attach your battery, the LED turns ON immediately because there is no voltage pressure pushing at the base, so the valve is in the opened position, allowing the current to flow from emitter to collector.

2. When you push the plunger down, the voltage immediately pushes against the base of the 3906 PNP transistor (Q1), closes the valve, and blocks the current flow. Voltage also fills the capacitor C1.

    After you release the push button, C1 holds the voltage pressure, and keeps voltage on the base, keeping the valve closed and current cut off.

3. As the voltage drains from C1 through R1, the voltage pressure against the base is released. The transistor starts passing current and voltage again slowly. The LED turns back on.

4. Why the extra resistor (R1): (a) before the push button is closed, both C1 and the base of the 3906 PNP transistor have NO voltage. Because there is no voltage pressure on the Q1's base, the valve is open and current flows from emitter to collector; (b) when the voltage in the capacitor is high, Q1's valve stays shut; (c) the path for current to escape from C1 through the transistor is blocked because the valve is closed; (d) so, R1 is necessary to drain the charge from the capacitor. This allows Q1's valve to open again.

    The capacitor is unable to drain and the transistor stays off because the voltage from the capacitor keeps the pressure on the base of the transistor, keeping the valve closed. The capacitor cannot drain through the base of the PNP transistor like it did in the previous 3904 NPN circuit. The extra resistor allows the cap to slowly drain, decreasing the voltage pressure on the base of the PNP transistor, allowing the valve to reopen and let current flow again.

Exercise: The PNP Transistor

1. In the schematic, Q stands for what component? Q represents the _____
    _____.

2. The arrow in the transistor symbol represents what action?

    a. Direction of current flow

    b. Direction of the collector

    c. Direction of the base

    d. Direction of the emitter

3. The arrow is always on the side of which leg in the schematic?

    a. Voltage

    b. Emitter

    c. Base

    d. Collector

4. a. What would happen if R3 were not in the circuit, and the LED was connected directly to the collector of the 3906 transistor?

    i.   LED would burn out.

    ii. LED would be bright.

    iii. LED would not work.

    iv. LED would flash.

    b. Explain your answer in 4a.

    _____

5. a. Replace C1 with the 100-microfarad capacitor. Describe what happens.

    _____

    b. Why does changing the capacitor affect the circuit this way?

    _____

6. Change C1 back to 10 microfarad. Now change R1 to 10 megohm (brown-black-blue). Describe what happens.

    _____

7. Think of the capacitor as a sink, holding water. Think of the resistor as the drainpipe. Which following statement best explains how changing to a higher resistance has the same effect as changing to a larger capacitor?

   a. The drain is bigger and empties the water faster.

   b. The drain is smaller and empties the water slower.

   c. The volume of water is bigger and takes longer to drain.

   d. The volume of water is smaller and drains faster.

8. a. Play a little. Replace R3 and the LED with the buzzer. Make sure the buzzer's red wire is getting voltage from the 3906's collector, and the black wire is connected to ground.

   b. Push the PB and release. What happens to the sound as the capacitor discharges.

   _____

9. Carefully describe in your own words how this circuit works.

   _____

Lesson 11: Your First Project: The Automatic Night Light

Build this circuit shown in Figure L11-1 on your breadboard. The photograph of the circuit appears in Figure L11-2 (see also Table L11-1).

Table L11-1 Parts list for Figure L11-2

Parts List	
D1	1N400X Diode
P1	100k Ω Pot
R1	22 kΩ
R2	470 Ω
LDR	1 MΩ dark
Q1	NPN 3904
LEDs	5 mm round

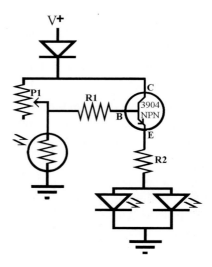

Figure L11-1

Figure L11-2

What to Expect

1. Attach your power supply.

2. Turn the knob on the pot one direction until the LEDs are barely off.

3. Now darken the room or stand in a closet. The LEDs will turn on.

4. Move to a semi-lit area. The LEDs will dim as you move into the light. Adjust the pot so the LEDs are again barely off. Any reduction of the amount of light will now turn the LEDs on. This is your automatic night light.

How It Works

Figure L11-3 shows the movement of current and voltage in the circuit as the light changes. Remember the NPN transistor needs positive voltage to its base to turn on.

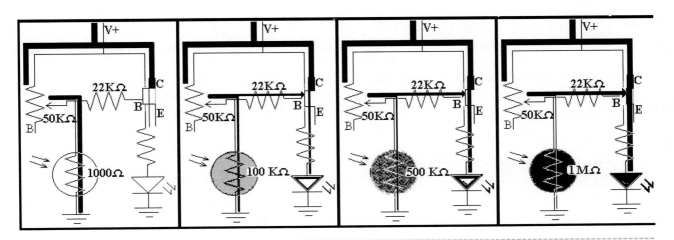

Figure L11-3

1. The potentiometer adjusts the amount of voltage shared by the 22-kilo-ohm resistor and the LDR.

2. In light, the LDR has low resistance, allowing all of the voltage to flow through to ground. Because the base of Q1 gets no voltage, the valve from C to E stays closed.

3. The resistance in the LDR increases as it gets darker, providing more voltage to the base of the transistor, pushing the valve open.

4. As the voltage flows through the transistor, the LEDs turn on.

If P1 is set to a low resistance, more voltage gets through. The more voltage that gets through the potentiometer, the easier Q1 turns on because the LDR cannot dump all of the voltage.

Exercise:
Your First Project—The Automatic Night Light

1. Set the pot so the resistance between legs "A" and "C" center is close to 50-kilo-ohm resistance. Figure L11-3 shows how the voltage and current move in the different conditions for this setup.

2. Now set your circuit under a fairly good light. Now measure and record the voltage at the test point shown in Figure L11-4. _____ V

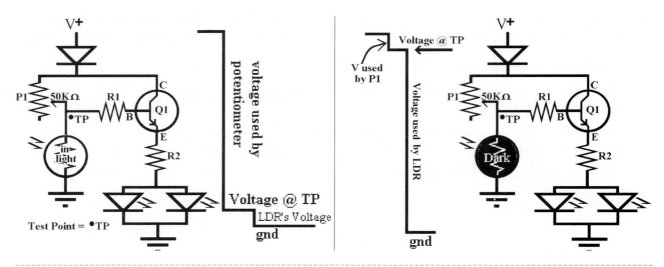

Figure L11-4

3. Now cover the LDR with a heavy dark pen cap. Measure and record the voltage at the test point again. _____ V

   Did the LED output change at all?

   So here's a major question?? Do you recall how much voltage is used from V+ to ground?

   • Does it depend on the amount of voltage available? NO!

   • Does it depend on the number or type of parts in the circuit? NO!

   By definition, voltage used between V+ and ground doesn't relate to any circuit variables.

The question is like asking, "How much distance is there between this altitude and sea level?"

The answer is—whatever the altitude is—or better said, all the distance!

Ground by definition is 0 volts.

• How much voltage is used between V+ and ground??

• The answer will always be *All the voltage is always used between V+ and ground*.

4. OK. Figure L11-5 will help explain how all the voltage in this circuit is used.

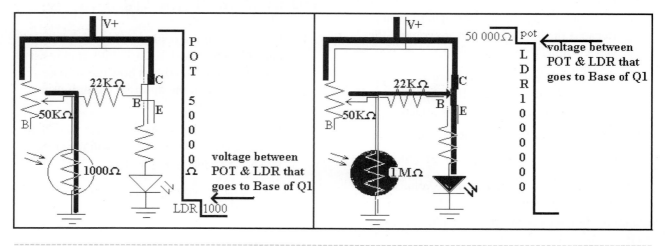

Figure L11-5

Remember the idea of the waterfall when we first looked at resistors. As the size of the load increased, the amount of voltage used increased *proportionately*. The same thing happens with two resistor loads in this circuit. More detail is given in Figure L11-6.

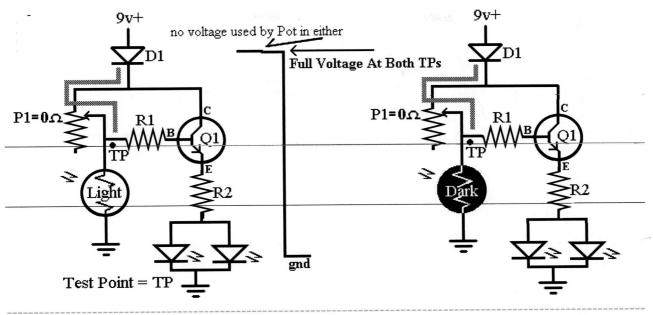

Figure L11-6

The pot uses some voltage because it is set to 50,000 ohm here.

The LDR uses a small amount of the voltage in the light because it has a small resistance. When it is in the dark, the LDR has a great deal of resistance. The base of the transistor reacts to the voltage available at that point where the LDR and pot connect. It becomes obvious which situation provides more voltage to the transistor's base.

5. Set the pot so the resistance between legs "A" and "C" center now has 0 ohm. Figure L11-6 shows how the voltage available to the transistor base is identical in both the light and the dark. The potentiometer uses none of the voltage, so the base of the transistor is exposed to nearly full voltage in both circumstances.

6. Now set your circuit under a fairly good light. Measure and record the voltage at the test point shown in Figure L11-6. _____ V

7. Now cover the LDR with a heavy dark pen cap. Measure and record the voltage at the test point again. _____ V

Did the LED output change at all? Did you expect the output to change?

8. Set the pot so the resistance between legs "A" and "C" center now has 100 kilo-ohm. Figure L11-7 shows how the voltage is responding to the changing resistance of the LDR, changing the voltage available to the base of the transistor. The pot here is adjusted to twice as much resistance as before, so it will use twice as much of the voltage available. The LDR must be set into nearly complete darkness or you will get muddy results.

9. Now set your circuit under a fairly good light. Measure and record the voltage at the test point. _____ V

10. Now cover the LDR with a heavy dark pen cap. Measure and record the voltage at the test point again. _____ V

Did the LED output change at all?

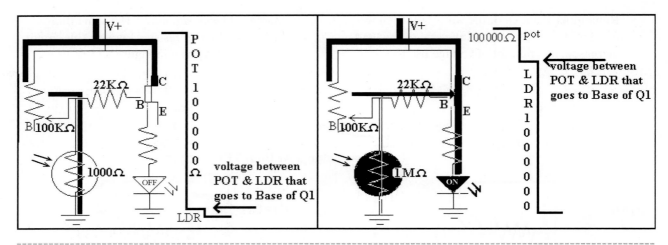

Figure L11-7

Building the Automatic Night Light Project

This is a top view of your *printed circuit board* (PCB). The traces on the underside are shown as gray in Figure L11-8.

Figure L11-9 shows the view of the PCB looking directly at the bottom. The copper traces replace the wires used in the solderless breadboard.

The series of frames in Figure L11-10 show how the PCB layout was developed. Creating such a layout is a relatively simple task.

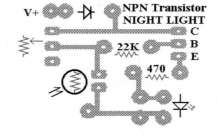

Figure L11-8

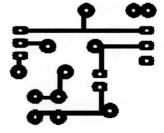

Figure L11-9

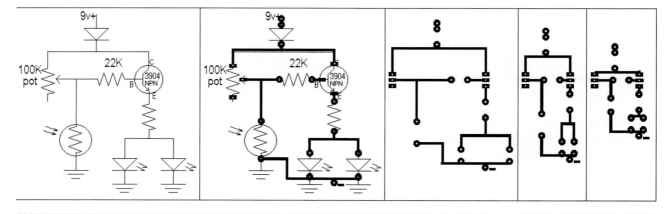

Figure L11-10

Mounting Your Parts

Be certain to get your parts into the correct holes as displayed in Figure L11-11. Your soldering technique is vital. Check out the soldering animation on the Web site at www.books.mcgraw-hill.com/authors/cutcher. Each solder connection should look like a Hershey's Kiss. Just as one bad apple spoils the bunch, one bad solder can spoil your project (and your fun). The parts are mounted onto the board so that their legs stick out through the bottom. The copper traces are on the bottom. The soldering is on the bottom. This is a good job.

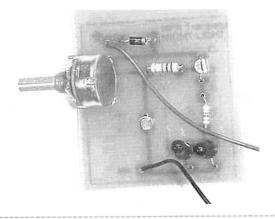

Figure L11-11

I got such a laugh when the circuit in Figure L11-12 was presented to me for troubleshooting. This person did not follow directions. They managed to mount the parts on the wrong side of the printed circuit board. They made a mess. It didn't work. There is never enough time to do it right, but there is always time to do it over.

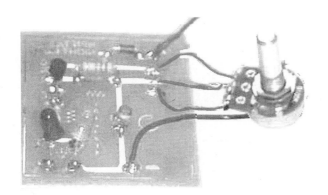

Figure L11-12

Finishing Up

I would recommend using a touch of hot glue on each corner of the PCB and then pressing the circuit onto a thick piece of cardboard. You can mount it on anything that is an insulating material. Anything metallic would short-circuit your project, and probably destroy the transistor.

Lesson 12: Specialized Transistors—The SCR

There are many highly specialized types of transistors. Here we use a latching switch called a *silicon-controlled rectifier* (SCR). This component is also referred to as a *Triac*. The SCR acts like a "trap door." Once triggered, it stays latched open. You are familiar with this because it is the basic component in fire alarms and burglar alarms. Once triggered, it stays on. This is your second project, a professional quality alarm circuit.

Electronics is all about using an electrical pulse to control things or pass information. A commonly used solid-state switch (no moving parts) is the SCR. It is provided in a variety of packages, the most common ones pictured next to their symbol in Figure L12-1.

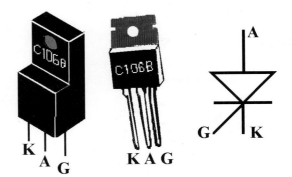

Figure L12-1

Remember, not everything in this type of package is an SCR.

"A" leg is anode, the positive side.

"G" leg is gate; not to be confused with ground, "gnd."

"K" leg is cathode, referring to the grounded side. "C" is already used for capacitor, e.g., C1, C2.

The SCR is often used in alarm systems; once it is triggered, it stays on. Its action is best depicted in the series of frames shown here in Figure L12-2.

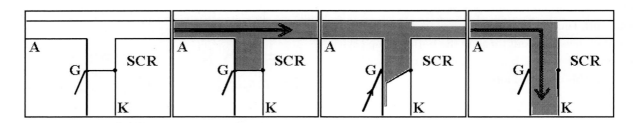

Figure L12-2

When V+ hits the G (gate) leg of the SCR, a latch releases, opening a "trap door" between A (anode) and K (cathode). This trap door remains open until the power is removed. In other words, it latches itself open. That is why an SCR is called a *latching circuit*. The only way to turn the SCR off is to shut off the power. Turn the power on again and the SCR is reset.

Breadboard the SCR Circuit

There are four stages that you need to take to build this professional style alarm circuit. Each one will be considered individually.

Stage 1: The Basic System

Build the basic SCR circuit by carefully following the schematic presented in Figure L12-3. The parts layout is shown in the photograph in Figure L12-4 (see also

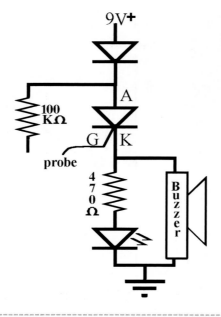

Figure L12-3

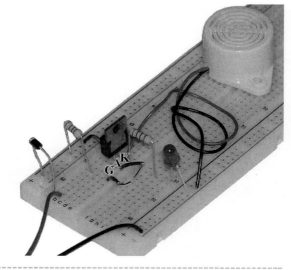

Figure L12-4

Table L12-1). Don't get too comfortable using the photograph. You will have to depend on schematics from now on to help you develop your alarm circuit.

Table L12-1   Parts list for Figure L12-4

Parts List

D1	1N400X
R1	100 kΩ
R2	470 Ω
SCR	Triac C106B
Buzzer	9 V
LED	5 mm round

What to Expect

When you attach the battery, it should be quiet.

Touch the sensor probe to the end of R1. The buzzer should turn on.

You need to disconnect the battery to reset the SCR.

How It Works

1. When you attach the battery, the LED is off and the buzzer is quiet. Voltage is not avail-
able to the LED and buzzer because the SCR has not been activated. The circuit path between A and K is not available until voltage is put to the gate.

2. When you touch the end of the probe to the bottom of the 100-kilo-ohm resistor, voltage is fed to the gate (G) leg.

3. The voltage activates the latch and opens the circuit path from A (V+) to K (gnd). Current moves on this path from A to K, through the SCR, providing voltage and power to the LED and the buzzer.

4. The buzzer and LED should turn on and stay on until you disconnect the power. When you reconnect the power, the LED should not light, and the buzzer should be quiet.

Stage 2: Make Life Easier

It is a hassle to disconnect the battery each time, just to reset the SCR. Wouldn't it be so much easier just to use a push button to reset the circuit? So go ahead and use the normally closed push button to do just that.

It does not matter where the voltage to the SCR is cut off. As you can see in Figure L12-5, any of the following suggestions work. Each one disconnects the voltage to the circuit and resets the SCR.

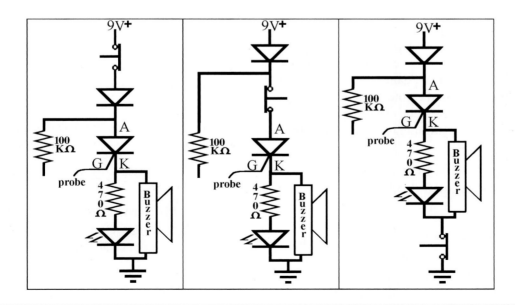

Figure L12-5

Stage 3: Avoid Static Buildup

Modify the SCR circuit with a few changes shown in Figure L12-6.

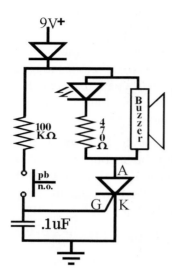

Figure L12-6

1. Shift the SCR below the LED and buzzer. In most circuits, you can be flexible. It really doesn't matter which component comes first.

2. Add the 0.1-microfarad capacitor.

3. Add the PBNO for a trigger.

How It Works

1. When the PB is pushed, the capacitor fills "nearly" instantly because the 100,000-ohm resistor slows down the current. The capacitor acts like a cushion and dampens any small but annoying jumps in the voltage to the gate that might be caused by static electricity. It avoids false alarms by preventing any accidental triggering of the SCR.

2. The gate (G) of the SCR senses the signal when the capacitor is filled.

Stage 4: A Complete System

Now modify your circuit to make a simple but professional quality alarm system by adding three more parts to your working breadboard as shown in Figure L12-7.

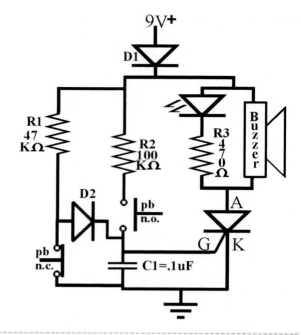

Figure L12-7

Table L12-2

Additional Parts

D2	1N400X
R1	47 kΩ
PBNC	Normally closed push button

What to Expect

Attach your power supply. The LED should be off and the buzzer quiet.

Push the plunger on the PBNC. The buzzer will start, and the LED light will turn on.

To reset, disconnect the power—count 10 seconds —and then reconnect.

Now push the plunger on PBNO. The buzzer and LED will turn on.

How It Works

As Long as PB1 stays closed, the voltage takes the "path of least resistance" and goes directly to ground. No voltage reaches G, the gate of the SCR.

When PB1 is pushed, the connection to ground is broken, and the voltage has to move through D2, activating G, turning the SCR on.

Lesson 12 — Specialized Transistors

As long as PB2 stays open, the voltage cannot move across the gap. Air is a great insulator. No voltage reaches G. When PB2 is pushed, the voltage travels through to the gate leg, turning the SCR on.

Exercise: Specialized Transistors — the SCR

1. If a component looks like your SCR, it is an SCR. True/false. Support your answer.

2. How can you tell if it is an SCR without putting onto your breadboard?

3. Many alarm systems use an SCR. For example, a fire alarm, once triggered, keeps going. Describe what needs to be done to reset the SCR in the alarm.

4. Use the schematic to follow the current's path when the PBNC is in its normal position.

5. How does the path change when you push the plunger on the PBNC?

6. Follow the current's path when the PBNO is in its normal position.

7. How does the path change when you push the plunger on the PBNO?

8. Have you ever seen metal foil tape on the edge of store windows? There is a small current running through this tape. If the window breaks, the foil is torn. What component does the foil tape replace in your burglar alarm system?

9. Can you describe how a normally open type of switch would be used in a burglar alarm?

Assembling the Alarm

Figure L12-8 represents the SCR's printed circuit board when viewed from the bottom. Figure L12-9 shows lines that appear faded. This is the view of the PCB when you look at the traces through the fiberglass backing.

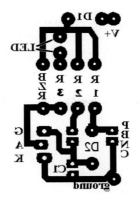

Figure L12-8

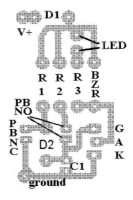

Figure L12-9

Consider this. If you mount your simple push buttons onto the PCB directly, you have a very cute demonstration circuit. If you mount the push buttons on a long wire, you still need to wait for someone to push one of the buttons. That still isn't much of an alarm. What you really need to do is create your own contact switches. When you do this, you can apply this as a real alarm system. The best setup for your power supply is to have both a wall adapter and a 9-volt battery. This way the circuit still works even if your house has no power.

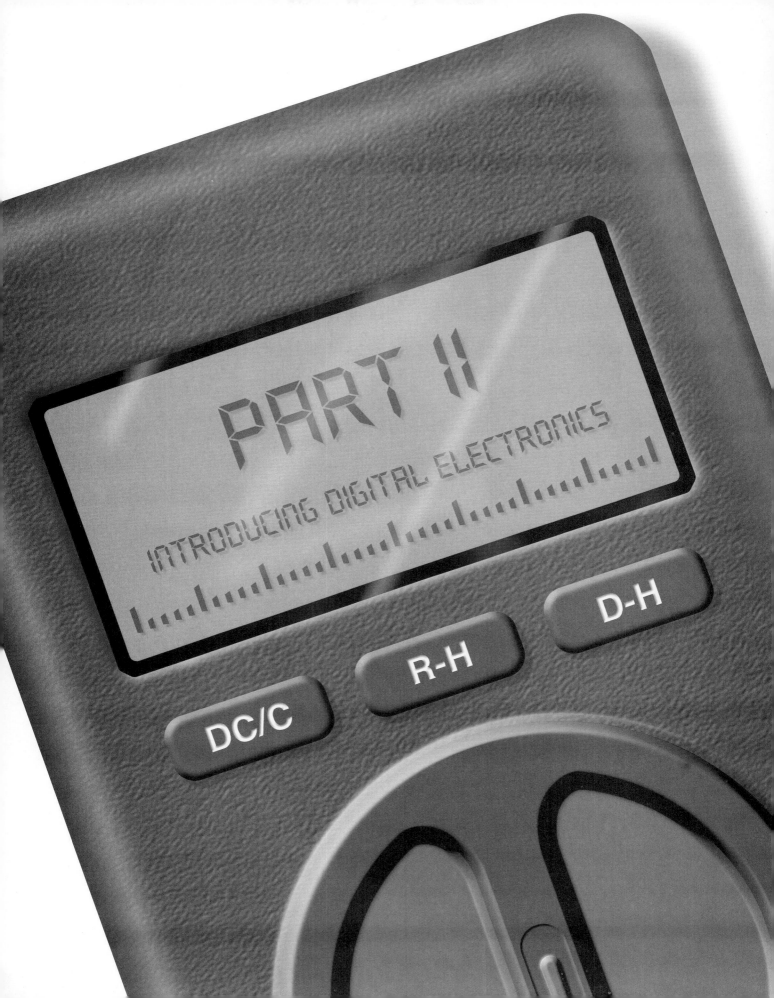

Digital Logic

Lesson 13: A Spoiled Billionaire

As we begin the "digital" electronics unit, this lesson page shows us a system that we can use to be certain all of the information is transferred all the time, perfectly. You must be able to count to 255 and know the difference between "on" and "off."

With apologies to Bill Gates but inspired by his explanation in his book, *The Road Ahead*.

There is an eccentric billionaire who lives near Seattle. He is particular about his lighting in different rooms of his homes, especially his den. He likes it set at exactly 187 watts. And when you're rich, you get what you like. But here's the real problem. His wife also uses the den and she "likes" 160 watts. They asked their groundskeeper to come up with a solution they could use in all their homes around the world. He first installed a dimmer switch and put a mark next to the spot that represented their preferences, which is pictured in Figure L13-1.

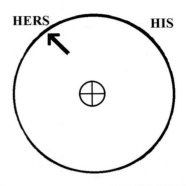

Figure L13-1

On inspection, the groundskeeper was told that his solution was not acceptable. It was not exact enough. So he thought further and had an idea. Actually, eight ideas.

His idea is shown in Figure L13-2. A light bar with eight separate specific value lights. Each light would have a different wattage rating, and each would have its own switch. To adjust the lighting for his own needs, the billionaire would only have to turn on a set of specific switches. Those are displayed specifically in Figure L13-3.

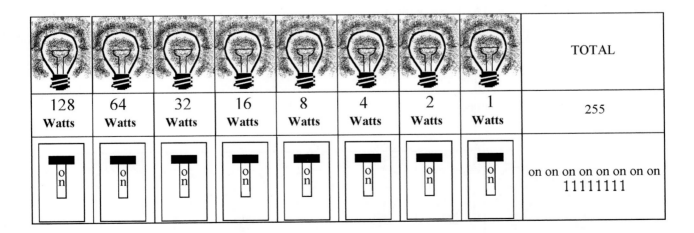

								TOTAL
128 **Watts**	64 **Watts**	32 **Watts**	16 **Watts**	8 **Watts**	4 **Watts**	2 **Watts**	1 **Watts**	255
on	on	on	on	on	on	on	on	on on on on on on on on 11111111

Figure L13-2

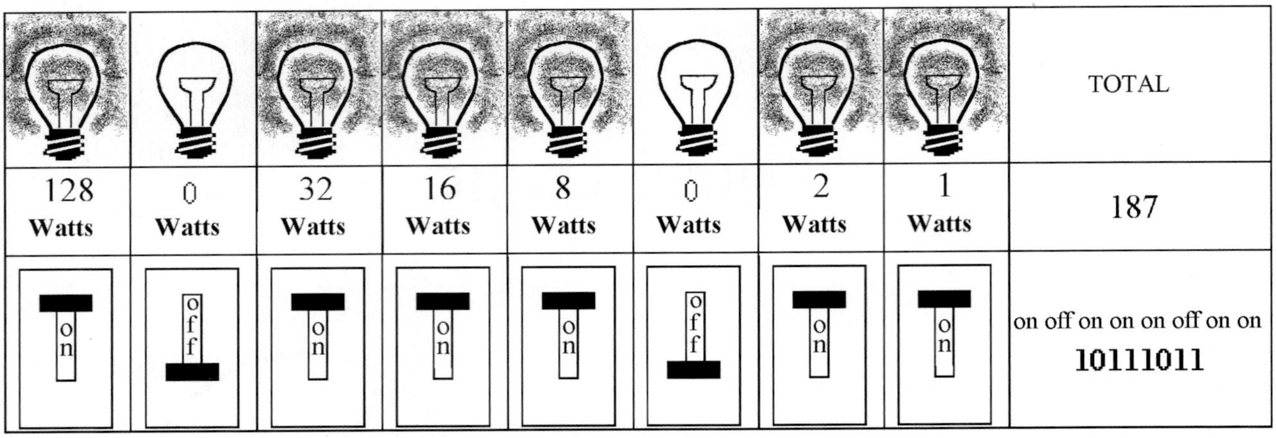

TOTAL
128 Watts
0 Watts
32 Watts
16 Watts
8 Watts
0 Watts
2 Watts
1 Watts
187
on off on on on off on on
10111011

Figure L13-3

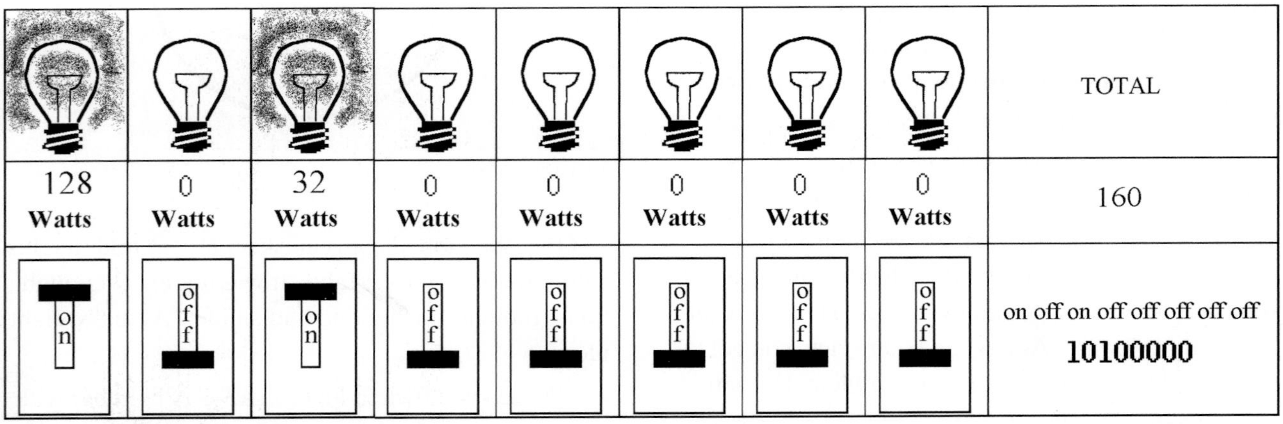

TOTAL
128 Watts
0 Watts
32 Watts
0 Watts
0 Watts
0 Watts
0 Watts
0 Watts
160
on off on off off off off off
10100000

Figure L13-4

And to make the softer lighting situation for Mrs. Billionaire, she flips the switches shown in Figure L13-4.

So all the groundskeeper had to do was write a note under the set of switches in the den.

His = 10111011

Hers = 10100000

(*The Road Ahead*, Bill Gates, pg. 25.)

"By turning these switches on and off, you can adjust the lighting level in watt increments from 0 watts (all switches off) to 255 watts (all switches on). This gives you 256 possibilities.

If you want 1 watt of light, you turn on only the rightmost switch, which turns on the 1-watt bulb (shown in Figure L13-5).

If you want 2 watts, you turn on only the 2-watt bulb (shown in Figure L13-6).

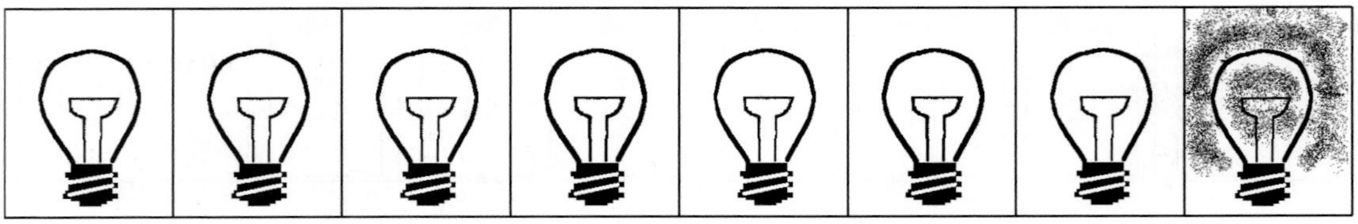

Figure L13-5

Figure L13-6

If you want 3 watts of light, you turn on both the 1-watt and 2-watt bulbs, because 1 plus 2 equals the desired 3 watts (shown in Figure L13-7).

If you want 4 watts of light, you turn on the 4-watt bulb.

If you want 5 watts, you turn on just the 4-watt and 1-watt bulbs (shown in Figure L13-8).

If you want 250 watts of light, you turn on all but the 4-watt and 1-watt bulbs (shown in Figure L13-9).

Figure L13-7

Figure L13-8

Figure L13-9

If you have decided the ideal illumination level for dining is 137 watts, you turn on the 128-, 8-, and 1-watt bulbs, like this (shown in Figure L13-10).

This system makes it easy to record an exact lighting level for later use or to communicate it to others who have the same light switch setup. Because the way we record binary information is universal—low number to the right, high number to the left, always doubling—you don't have to write the values of the bulbs. You simply record the pattern of switches: On,

off, off, off, on, off, off, on. With that information a friend can faithfully reproduce the 137 watts of light in your room. In fact, as long as everyone involved double-checks the accuracy of what he does, the message can be passed through a million hands; and at the end, every person will have the same information and be able to achieve exactly 137 watts of light.

To shorten the notation further, you can record each "off" as 0 and each "on" as 1. This means that instead of writing down "on, off, off, off, on, off, off, on," meaning turn on the first, the fourth, and the eighth of eight bulbs and leave the others off, you write the same information as 1,0,0,0,1,0,0,1, or 10001001, a binary number. In this case, it's 137. You call your friend and say: "I've got the perfect lighting level! It's 10001001. Try it." Your friend gets it exactly right, by flipping a switch on for each 1 and off for each 0.

This may seem like a complicated way to describe the brightness of a light source, but it is an example of the theory behind binary expression, the basis of all modern computers.

The simplest computers use an 8-bit system like the 8 light switches. Each bit is a bit of information. The binary word of 8 bits makes 1 byte, as shown in the explanations above.

From the earliest days of computing, the alphabet and numerals have been assigned specific values.

Table L13-1   The binary alphabet: The ASCII table

Space = 20 = 00010100

A = 65 = 01000001 a = 97 = 01100001

B = 66 = 01000010 b = 98 = 01100010

C = 67 = 01000011 c = 99 = 01100011

D = 68 = 01000100 d = 100 = 01100100

E = 69 = 01000101 e = 101 = 01100101

F = 70 = 01000110 f = 102 = 01100110

G = 71 = 01000111 g = 103 = 01100111

H = 72 = 01001000 h = 104 = 01101000

I = 73 = 01001001 i = 105 = 01101001

J = 74 = 01001100 j = 106 = 01101010

K = 75 = 01001101 k = 107 = 01101011

L = 76 = 01001110 l = 108 = 01101100

M = 77 = 01001101 m = 109 = 01101101

N = 78 = 01001110 n = 110 = 01101110

O = 79 = 01001111 o = 111 = 01101111

P = 80 = 01010000 p = 112 = 01110000

Q = 81 = 01010001 q = 113 = 01110001

R = 82 = 01010010 r = 114 = 01110010

S = 83 = 01010011 s = 115 = 01110011

T = 84 = 01010100 t = 116 = 01110100

U = 85 = 01010101 u = 117 = 01110101

V = 86 = 01010110 v = 118 = 01110110

W = 87 = 01010111 w = 119 = 01110111

X = 88 = 01011000 x = 120 = 01111000

Y = 89 = 01011001 y = 121 = 01111001

Z = 90 = 01011100 z = 122 = 01111010

The real question for the rest of us becomes— Why use binary? It seems so confusing!

The answer is actually very simple. We are dealing with machines. The easiest thing for a machine to sense is whether something is on or off.

So we are forced to use a system that can count in ons and offs.

But still you ask, why bother?

Well, look at it (Table L13-2).

Table L13-2 Compare analog to digital

	Analog	*Digital*
Advantages	1. Varying voltages.	1. Precise transfer of information.
	2. Easy to record.	2. No generational loss.
	3. Easy to play back.	3. Footprint of a bit can be done at the molecular level.
		4. Common transfer rate in billion bits per second.
		5. Any material can be used for storing data: 0s and 1s.
Disadvantages	1. Not precise.	1. Needs special equipment to transfer, record, and read information.
	2. Signal loss with each generation recorded. Have you ever watched a copy of a copy of a video tape? Ugh!	
	3. Takes up large "recording" space. Compare the size of a video tape to a DVD or a Memory Stick.	
	4. Limited transfer time—how long does it take to record a video, compared to transfer of a CD or MP3 file?	
	5. Specific media used for storage doesn't translate from one media to another. For example, video to film.	
	6. Signal fades as media ages.	

Exercise: A Spoiled Billionaire

I am writing in binary code!

```
0100100100010100011000010110110100010100011
1101110111001001101001011001110110100001111
0100011010010110111001101110001010001101000
1011011100001010001100010011010010110110100
1100001011100100111100100010100011000011011
0111101100100011001001
```

1. For yourself, what is the most important advantage of using digital information.

   Use this chart to help translate from binary to decimal, and back.

Bit number	Bit 8	Bit 7	Bit 6	Bit 5	Bit 4	Bit 3	Bit 2	Bit 1
Value	128	64	32	16	8	4	2	1

2. Translate the following 8-bit binary codes to the decimal equivalent.

Binary	Decimal
10101100	
01100110	
10010011	
00110001	

3. Translate the following decimal numbers to the binary codes.

Binary	Decimal
	241
	27
	191
	192

4. Consider this. If 8 bits count up to 255 decimal, what is the maximum that a 9-bit binary word can count?

Lesson 14: The Basic Digital Logic Gates

This course deals with a vast amount of new information. It is not hard. Just new.

You learn about the five main types of logic gates in this lesson. Each gate is built with individual transistors that act like normally open or normally closed push buttons. All these switches do is redirect the output to voltage (V) or ground.

The AND gate and NOT AND gate are perfectly opposite in their outputs.

The OR gate and NOT OR gate are also perfectly opposite in their output.

Table L14-1 In digital electronics, the *voltage state* is named in three different ways

Voltage States

V+	Ground
1	0
High	Low

- Each term for voltage has an alternate name for ground.
- These terms are generally interchangeable.
- The terms are usually paired.

- Terms like V+ and ground are used together just like the term *high* is used with *low*, 1 with 0.
- Digital gives an *output* of **high** or **low**, but we often have to use a real world *analog input*.

All gates have at least 1 input, but all gates have only 1 output.

Inputs are the analog "sensors." They compare the voltage they feel to the voltage of the chip. They sense if the input is to be seen as high or low.

Output is the result of the logic function, whether the gate provides a full V+ or ground at the output.

1. This real world input of analog is usually never given to us as a convenient high or low.
2. So Digital gates have been designed to compare the input against their own power source of V+.
3. Anything above 1/2 of V+ is seen as a high input.
4. Anything below 1/2 of V+ is seen as a low input.

Inputs Are the Analog Sensors

- For example, as shown in Figure L14-1, the gate is powered by 10 volts.
- Anything connected to the input that is over 5 volts is seen as a high input.
- Alternatively, any input voltage that is below 5 volts is seen as low input.

The chip contains inputs, processors, and outputs.

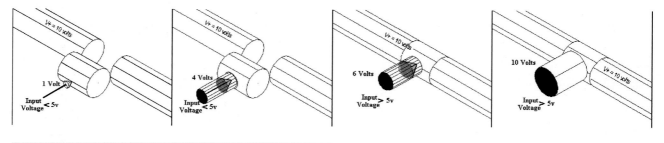

Figure L14-1

Each of these need power. They are all powered from the same power source. The voltage source is shown as the line coming into the symbol marked with the V+.

The NOT Logic Gate

The input given is NOT the output. You can see this in Figure L14-2.

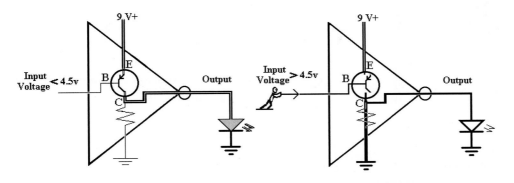

Figure L14-2

This is often referred to as an inverter. It "inverts" the input.

- You can see that the input is like a push button. It controls the voltage flow through the transistor switch inside the chip.

- The transistor processor inside is like the push-button demonstration circuits. The transistor responds to the input, controlling voltage to 1the output.

For the purpose of this exercise, refer to Figure L14-3.

- The pushed button is high.

- The unpushed button is low.

Breadboard the NOT Gate Simulation Circuit

This circuit quickly demonstrates the action of the NOT gate (see Figure L14-4).

The input is the force of your finger. The binary processor is the push button's position (up = 0, down = 1). The output is the LED. On is high, and Off is low (Table 14-1).

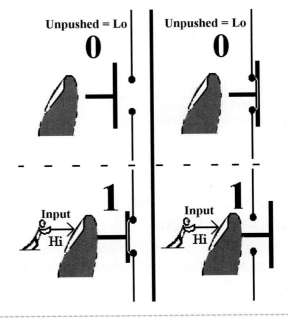

Figure L14-3

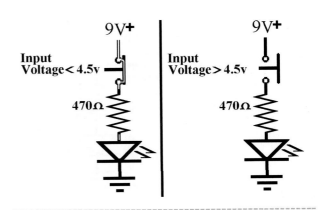

Figure L14-4

Table 14-1 NOT gate: Complete the logic table for this gate

Input	Output
High	___
Low	___

The AND Logic Gate

As shown in Figure L14-5, both input A AND input B have to be HI to get a HI output.

These inputs are like normally open push buttons. They control the voltage flow through the NPN transistors inside the chip. The transistor processors inside are like the push-button demonstration circuits below. The transistors respond to the inputs, controlling voltage to the output.

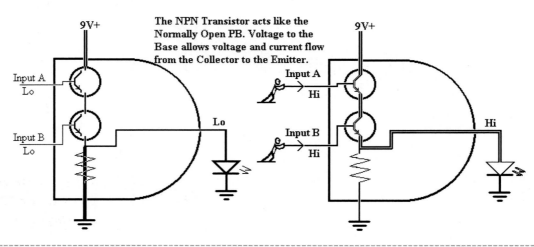

The NPN Transistor acts like the Normally Open PB. Voltage to the Base allows voltage and current flow from the Collector to the Emitter.

Figure L14-5

Breadboard the AND Gate Simulation Circuit

Build the simulation circuit that demonstrates the action of the AND gate as displayed in Figure L14-6. Remember that for the purpose of this exercise, as shown in Figure L14-3:

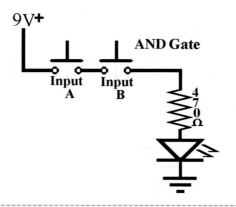

Figure L14-6

- The pushed button is high.
- The unpushed button is low.

The NPN transistors act like the normally open push buttons. Voltage at the base greater than 1/2 of V+ allows the voltage to move through (Table L14-2).

Table L14-2 AND gate: Complete the logic table

Input A	Input B	Output
High	High	___
High	Low	___
Low	High	___
Low	Low	___

The OR Logic Gate

As shown here in Figure L14-7, input A OR input B has to be HI to get a HI output. The inputs still act like normally open push buttons. They control the voltage flow through the NPN transistors inside the chip. The transistor processors inside are like the push-button demonstration circuit below. The transistors respond to the input, controlling the voltage to the output.

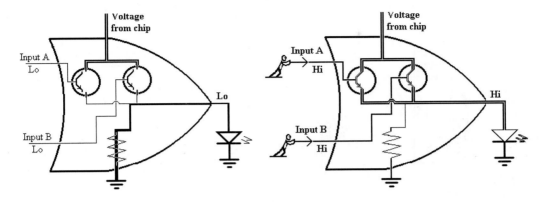

Figure L14-7

Breadboard the OR Gate Simulation Circuit

Build the simulation circuit that demonstrates the action of the OR gate as displayed in Figure L14-8.

Remember that for the purpose of this exercise, as shown in Figure L14-8:

- The pushed button is high.
- The unpushed button is low.

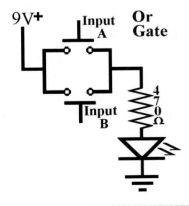

Figure L14-8

The NPN transistors act like the normally open push buttons. Voltage at the base greater than 1/2 of V+ allows the voltage to move through (Table L14-3).

Table L14-3 OR gate: Complete the logic table

Input A	Input B	Output
High	High	____
High	Low	____
Low	High	____
Low	Low	____

The NAND Logic Gate

As shown here in Figure L14-9, the NAND gate looks a bit more complex. It really isn't. It is designed to give the exact opposite results of an AND gate. That is why it is referred to as a NOT AND gate.

The inputs act like normally closed push buttons. They control the voltage flow through the PNP transistors inside the chip. The transistor processors inside are like the push-button demonstration circuit below. The transistors respond to the input, controlling the voltage to the output.

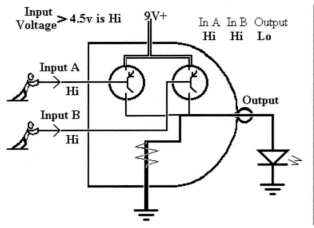
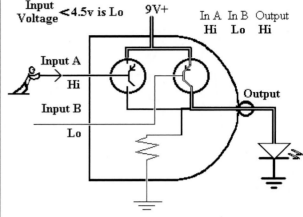

Figure L14-9

Breadboard the NAND Gate Simulation Circuit

Build the simulation circuit that demonstrates the action of the NAND gate as displayed in Figure L14-10.

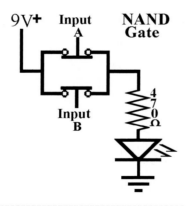

Figure L14-10

The PNP transistors act like normally closed push buttons. Voltage at the base greater than 1/2 of V+ stops the voltage from moving through (Table L14-4).

Table L14-4 NAND gate: Complete the logic table

Input A	Input B	Output
High	High	____
High	Low	____
Low	High	____
Low	Low	____

The NOR Logic Gate

The NOR gate is displayed in Figure L14-11. Just like the NAND gate, it looks overly complex. Relax. It, too, was designed to give the exact opposite results of an OR gate. That is why it is referred to as a NOT OR gate. The inputs act like normally closed push buttons. They control the voltage flow through the PNP transistors inside the chip. The transistor processors inside are like the push-button demonstration circuit below. The transistors respond to the input, controlling the voltage to the output.

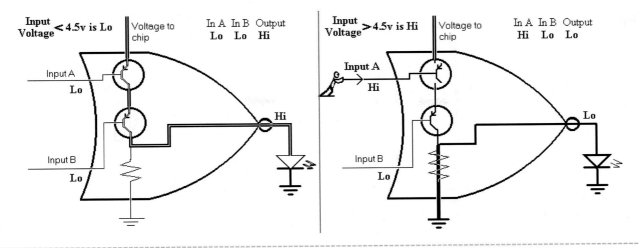

Figure L14-11

Breadboard the NOR Gate Simulation Circuit

Build the simulation circuit that demonstrates the action of the NOR gate as displayed in Figure L14-12.

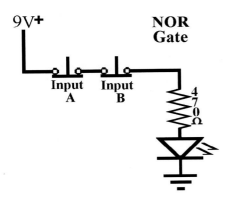

Figure L14-12

The PNP transistors act like normally closed push buttons. Voltage at the base greater than 1/2 of V+ stops the voltage from moving through (Tables L14-5 and L14-6).

Table L14-5 NAND gate: Complete the logic table

Input A	Input B	Output
High	High	___
High	Low	___
Low	High	___
Low	Low	___

Table L14-6 Comparing the gates

NOT		AND			OR			NAND			NOR		
In	Out Put	In A	In B	Out Put	In A	In B	Out Put	In A	In B	Out Put	In A	In B	Out
High	Low	High	High	High	High	High	High	High	High	Low	High	High	Low
Low	High	High	Low	Low	High	Low	High	High	Low	High	High	Low	Low
		Low	High	Low	Low	High	High	Low	High	High	Low	High	Low
		Low	Low	Low	Low	Low	Low	Low	Low	High	Low	Low	High

1. What did you use to represent the inputs?

2. What component is used to represent the processor?

3. What component is used to represent the output?

4. Where does the voltage powering the output come from?

5. When the push buttons were *un*pushed, that represented a state of high or low at the input?

6. When the push buttons were pushed, that represented a state of high or low at the input? Look again at the graphics of the logic gates.

7. What components are actually used in the IC chip as the processors?

8. The input is actually similar to which part of the transistor?

Lesson 15: Integrated Circuits CMOS ICs

There are thousands of *integrated circuits* (ICs). The 4000 series of CMOS ICs are very popular because they are inexpensive, and work with as little as 3 volts and as much as 18 volts. With mishandling, they are easily destroyed with static electricity. So they go from packing material to your SBB. Some basic vocabulary is developed. Common layout for these chips is discussed. Handling instructions are given. And mentioned again. Did I already say that these things are static sensitive? Don't rub them in your hair. They won't stick to the ceiling like balloons afterward. They just won't work.

You need this information because in the next chapter, you will:

- Build a prototype of a digital alarm system

- Learn to use a variety of events that can start the system

- Learn how to determine the system output

- Learn to determine how long the system stays on after it is triggered.

The IC that you will use is the 4011 CMOS. The 4011 chip is set into a 14-pin *dual inline package* (DIP). Figure L15-1 shows an 18-pin chip in a DIP format.

Figure L15-1

Precautions

The 4000 series CMOS IC has been used in electronics since the '70s. They are versatile and widely used. They are used here because they work on a range from 3 to 18 volts. They are inexpensive, too! **But they are static sensitive.** You know. Shuffle across the carpet and zap your friend. Even the smallest zap can toast a CMOS chip (see Figure L15-2).

Ignore these precautions at your own risk!

1. Always store the IC in a carrying tube or in "static" foam until it is placed into the circuit.

2. Remove static from your fingers. Touch some type of large metal object to remove any static electricity from your fingers before you handle the CMOS chips.

Figure L15-2

3. Don't walk across the room with a CMOS chip in hand. Walking across linoleum or a rug in a dry room will build up a static charge.

4. Always check that the chip is set in properly. A backward chip is a dead chip ($$).

5. Always tie any unused inputs to ground. I'll note where we have done that in this circuit.

6. Having an unused input pin unconnected is not the same as tying it to ground. If an input is not connected, the small voltage changes in the air around us can affect the input.

Take a Moment and Look at the Partial List of CMOS Series

For a more complete idea of the chips available, visit www.abra-electronics.com.

4000 Dual 3-input NOR gate plus inverter

4001 Quad 2-input NOR gate

4002 Dual 4-input NOR gate (same as 74HC4002)

4006 18-stage shift register, serial-in/serial-out

4007 Dual CMOS pair plus inverter

4008 4-bit, full-adder arithmetic unit

4009 Hex inverter—OBSOLETE, use 4049 instead

4010 Hex buffer—OBSOLETE, use 4050 instead

4011 Quad 2-input NAND gate

4012 Dual 4-input NAND gate

4013 Dual Type D Flip-Flop

4014 8-stage shift register, parallel-in/serial-out

4015 Dual 4-stage shift register, serial-in/parallel-out (= 74HC4015)

4016 Quad bilateral analog switch

4017 Decade counter, synchronous 1-of-10 outputs

4018 Programmable counter, walking ring

4019 4-pole, double-throw data selector

4020 14-stage binary ripple counter (same as 74HC4020)

4021 8-stage shift register, parallel-in/serial-out

4022 Octal counter, synchronous 1-of-8 outputs

4023 Triple 3-input NAND gate

4024 7-stage binary ripple counter (same as 74HC4024)

4025 Triple 3-input NOR gate

4026 Decade counter and 7-segment decoder with enable

4027 Dual JK Flip-Flop with preset and clear

4028 1-of-10 decoder

4029 Up-down synchronous counter, decade or hexadecimal

4030 Quake EXCLUSIVE-OR gate— OBSOLETE, use 4077 or 4507

4031 64-stage shift register, serial-in/serial-out

4032 Triple-adder, positive-logic arithmetic unit

4033 Decade counter and 7-segment decoder with blanking

4034 8-bit bidirectional storage register

4035 4-stage shift register, parallel-in/parallel-out

4038 Triple-adder, negative-logic arithemtic unit

4040 12-stage binary ripple counter (same as 74HC4040)

4041 Quad inverting/noninverting buffer

4042 Quad latch storage register

4043 Quad Flip-Flop, R/S NOR logic

4044 Quad Flip-Flop, R/S NAND logic

4046 Phase-locked loop, special device

4047 Astable and monostable multivibrator

4049 Hex inverter/translator (same as 74HC4049)

4050 Hex buffer/translator (same as 74HC4050)

4051 1-of-8 analog switch (same as 74HC4051)

4052 Dual 1-of-4 analog switch (same as 74HC4052)

4053 Triple 1-of-2 analog switch (same as 74HC4053)

4060 14-stage binary ripple counter with oscillator (= 74HC4060)

4063 4-bit magnitude comparator arithmetic unit

4066 Quad analog switch, low-impedance (same as 74HC4066)

4067 1-of-16 analog switch

4068 8-input NAND gate

4069 Hex inverter

4070 Quad EXCLUSIVE-OR gate

4071 Quad 2-input OR gate

4072 Dual 4-input OR gate

4073 Triple 3-input AND gate

4075 Triple 3-input OR gate (same as 74HC4075)

4076 4-stage tri-state storage register

4077 Quad 2-input EXCLUSIVE-NOR gate

4078 8-input NOR gate (same as 74HC4078)

4081 Quad 2-input AND gate

4082 Dual 4-input AND gate

4089 Binary rate multiplier, special device

4093 Quad 2-input NAND Schmitt Trigger

4097 Dual 1-of-8 analog switch

The 4011 Dual Input Quad NAND Gate

The 4011 IC is a semiconductor that looks like Figure L15-3.

Figure L15-3

Notice the method of numbering the pins. All DIPs use this system.

Looking at the IC from the top, the reference notch should be to the left as shown. Then, pin 1 is on the bottom left. The numbering starts there and moves counterclockwise.

What happens when you flip the chip upside down? Where is pin 1 now?

Study Figure L15-4. The 4011 IC contains four separate NAND gates, each able to work independently. It has 14 pins. These pins are the connecting points.

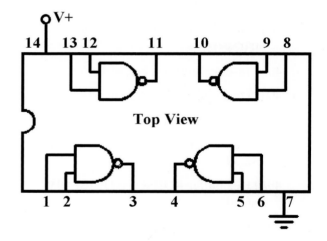

Figure L15-4

Here is the pin-out diagram for the 4011.

- Each pin has a specific function for each IC. It is important to connect these pins correctly.

- Pins 1 and 2 are inputs to the NAND gate that has its output at pin 3.

- Power to the chip is provided through pin 14.

- The chip's connection to ground is through pin 7.

- The inputs redirect the output of each gate to voltage (pin 14) or to ground (pin 7).

Exercise: Integrated Circuits, CMOS ICs

1. What is a DIP? _____

2. State the functional voltage range of a CMOS IC. _____ to _____ V

3. Briefly state the six major precautions regarding the proper care and feeding of an IC.

   a. _____

   b. _____

   c. _____

   d _____

   e. _____

   f. _____

4. Draw a picture of your 4011 chip.

5. Indicate on your drawing all writing on the chip.

6. Include the marker notch on your drawing.

7. Label pins 1 through 14.

8. From the diagram sheet, which pin powers the 4011 chip? _____

9. Which pin connects the chip to ground?
   _____

10. Describe clearly how to identify pin 1 on any IC. _____

11. What would happen if a wire connection of the 4011 was made to the wrong pin?

    _____

    _____

12. Would your answer for the previous question be true for any IC? Look at the list of some CMOS ICs mostly the 4000 series.

13. Look at the list of some of the CMOS ICs of the 4000 series. How many ICs shown are dedicated Logic Gate chips?

The First NAND Gate Circuit

In this chapter we will do the following:

- We will build and become familiar with a basic digital circuit.

- We will learn how an analog signal is translated into a digital output, using different input devices.

- We will be introduced to control timing in a resistor-capacitor circuit.

Lesson 16: Building the First NAND Gate Circuit

Enough talk, already. Back to the fun stuff.

Breadboard the circuit shown in Figures L16-1 and L16-2 (see Table L16-1).

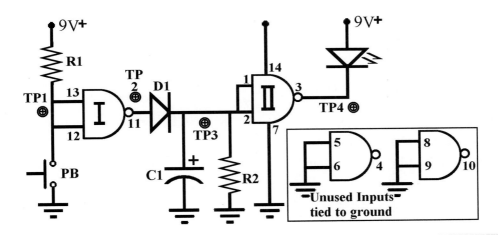

Figure L16-1

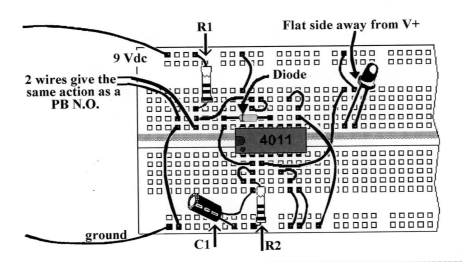

Figure L16-2

Table L16-1

Parts List	
R1	100 kΩ
R2	10 MΩ
C1	1 µF axial or radial
LED	5 mm round
D1 *	Signal diode (skinny golden)
IC 1	4011 Quad NAND gate

*Don't use the fat golden diode. You can substitute the black power diode.

Even though they don't look the same, match the pin numbers in the drawing with the pin numbers on the schematic. Refer back to the 4011 pin-out diagram of Figure L15-5 to help you with this.

What to Expect

When you attach the power supply, the LED should stay off.

The circuit is working when you momentarily close the PBNO and the LED turns on for about 8 seconds, and then automatically turns off again.

If the LED turns on as soon as you attach the power supply, immediately disconnect the power. Something is wrong. If the power remains connected, you could burn out your chip. Also, if all you wanted to do was turn an LED on, what's all the other stuff doing in the circuit? Just use an LED and resistor for that.

If the circuit refuses to work immediately, you need to refer to the Troubleshooting Section.

Troubleshooting

Some general questions to ask as you look for errors.

1. Is the power connected properly? Are you positive about that? Check again!

2. Are your breadboard connections done properly? All connections need to be in the small rows of five dots. Look quickly for any side-by-side connections.

3. Are your parts in the right way?
   - Is your chip in the right way?
   - Is your diode going the right way, as shown in the schematic?
   - A backward capacitor won't help either.
   - A backward LED won't turn on, even if everything else is working.

4. Now examine Figure L16-3.

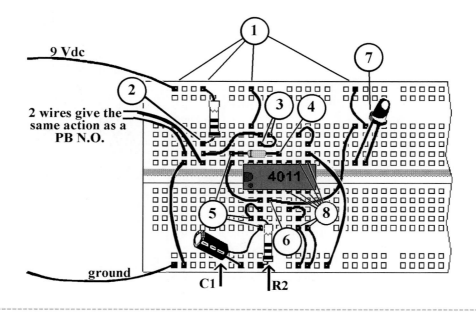

Figure L16-3

- Set your DMM to voltage DC. Connect the black probe to ground.

- Use the red probe to take two measurements of the voltage at each of the 19 check points. The first measurement is when the circuit is at rest. That is, you have connected the power supply but not played with the push button. Record your results.

- The second set of measurements is when the push button is being held down, and pins 12 and 13 are being connected to ground through the push button. This is to guarantee that you are measuring the circuit in its active state. See Table L16-2.

Focus on the area that does not match the expected results.

Table L16-2 Measurements for Figure L16-3

	At Rest	Active
		Pins 12 and 13 Connected to Ground
1.	V+ These points are connected directly to voltage.	V+ These points are still connected directly to voltage.
2.	This is a bit less than V+.	0 Volts because when the PB is closed, it is connected directly to ground.
3.	This is the same as 2.	This is the same as 2.
4.	Reading should be 0.0 v.	Reading should be V+.
5.	0 Volts or close to it.	Reading should be V+.
6.	The readout from pin 3 should read V+.	This should be V+ minus 2 V (LED uses 2 V) Remove LED and reading should change to V+.
7.	LED Flat side toward pin 3 LED should be off.	LED should be On, remains on for 8 or so seconds after you release the push button.
8.	Pins 5, 6, 8, and 9 should have 0.0 V.	These are connected directly to ground and should still have 0.0 V.

Lesson 17: Testing the Input at Test Point 1

If the push button works, you will measure the voltage when it is not pushed. There will be no voltage when you push it because you are connecting directly to ground when it is pushed.

Record all of your results on the data table at the end of Section Five.

Wrap a piece of wire around the common probe (black) and attach it to the ground line of your breadboard. Figure L17-1 shows clearly how to do it. This makes it easier to do all the other things you need to do as you use the red probe to measure the voltage at different test points.

Figure L17-1

1. Attach the power supply and measure the voltage at the test point 1 (TP 1 shown in Figure L16-1).

What Is Happening Here?

- Figure L17-2 is showing that the voltage through R1 is putting pressure on pins 12 and 13, the connected pair of inputs to one of the 4011 NAND gates. The input voltage is seen as a high.

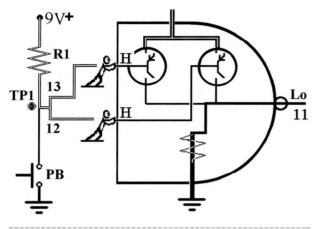

Figure L17-2

- When you measure the voltage at TP1, without pressing the normally open push button, you are measuring the actual voltage at the input to the NAND gate.

  Is that voltage greater than half of the power source voltage? _____

  It should be much greater than half of the voltage.

  Describe the *state* of the inputs to the first NAND gate when the system is at rest._____

2. With the power supply still connected, push the plunger of the push button down. Record the voltage.

What Is Happening Here?

- Figure L17-3 shows that when you close the push button, the voltage through R1 flows directly to ground. Like water, the voltage and current always takes the easiest path.

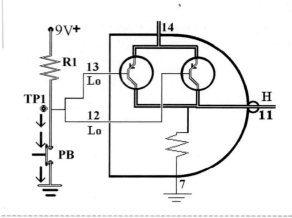

Figure L17-3

- The connected pair of inputs 12 and 13 no longer have any voltage pressure voltage on them. They are all connected to ground. Ground at 0.0 volts is definitely less than half of the voltage supplied to the chip. Remember that ground is just another word for a low state. Is that voltage greater than half of the power source voltage?

  _____

  It should read 0.0 volts. With the plunger down, it should be connected directly to ground.

  Describe the *state* of the inputs to the first NAND gate when the push button is closed.

  Describe what happens to the voltage on the DMM when you release the push button.

  The other instrument reading shown in Figures L17-4 and L17-5 is an oscilloscope. Each horizontal line equals 2 volts. The line representing voltage is at 4 1/2 lines up. 4.5 × 2 represents 9 volts.

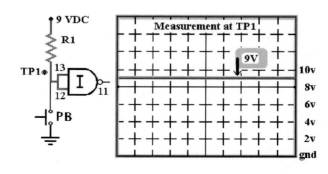

Figure L17-4

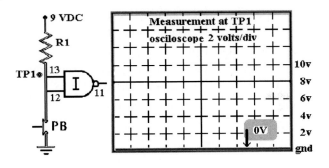

Figure L17-5

Don't worry if you don't have a proper oscilloscope available. Your computer can be used as a cheap "scope" with freeware available online. Winscope 2.51 was developed by Konstantin Zeldovich, Ph.D.

It is a wonderful tool and available at http://polly.phys.msu.su/%7Ezeld/index.html. The only failing is that it is limited to measuring rapidly changing signals. We'll be doing that soon enough. It will not measure a stable DC input.

Lesson 18: Test Point 2— The NAND Gate Processor at Work

For the NAND gate, if the inputs are connected to voltage, the output should be connected to ground. Conversely, when the inputs are connected to ground, the outputs are connected to voltage. The results happen instantly. Take a minute and understand the NAND gate.

What to Expect

With the system at rest and the push button untouched, measure the voltage at the output of the first NAND gate at pin 11 (Table L18-1).

Table L18-1   NAND gate logic table

Input A	Input B	Output
High	High	Low
High	Low	High
Low	High	High
Low	Low	High

The measurement for TP 2 is at pin 11. Referring to the schematic in Figure L18-1, this is right at the output of the first gate, but before the diode.

Record all of your results on the data table at the end of Section Five.

1. Measure the voltage at TP 2 with the push button open and

   - Because pins 12 and 13, the inputs to the NAND gate, are high, the output at pin 11 is low.

   - This would be the expected value of the output when the circuit is "at rest."

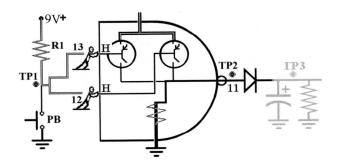

Figure L18-1

2. Measure the voltage at TP 2 with the push button held closed and record your results.

   - Both inputs of the NAND gate are now connected directly to ground; that is to say, they are **tied to ground** as shown in Figure L18-2. Remember that in the NAND gate, if either input receives a low produces a high output.

   - This would be the expected value of the output when the circuit is just started, or "activated."

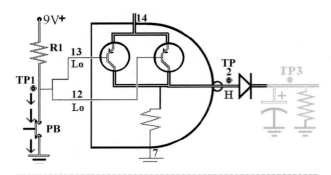

Figure L18-2

The output at pin 11 is normally low when the circuit is at rest. Remember low equals ground.

This is how the output at TP 2 would look on an oscilloscope. Figure L18-3 shows where the voltage would be when the circuit is at rest. Figure L18-4 shows where the voltage would be when the plunger is held down and pin 11 is connected to ground.

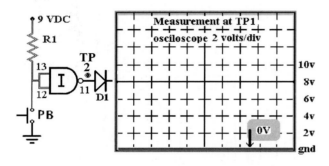

Figure L18-3

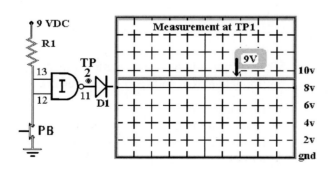

Figure L18-4

The voltage produced at pin 11 pushes through the diode and gets trapped on the other side. Referring to the schematic in Figure L18-4, it would be OK to say the voltage and current are trapped on the right-hand side of the diode, and not able to drain to ground through pin 11. It reaches ground another way.

Lesson 19: Test Point 3— Introducing the Resistor Capacitor Circuit

You are introduced to a resistor/capacitor working as a pair to control timing. You saw this briefly when you were introduced to capacitors in Part I. This is called an *RC Circuit*. This is actually one of the major subsystems in electronics, used to control timing. Here it is used to control how quickly the voltage to the second NAND gate drains, which controls how long the LED stays on. In an RC timing circuit, imagine this. The capacitor is like a sink and holds the charge. The resistor is like a drain pipe. It allows the charge in the sink to flow out.

Note that TP 3 can be considered anywhere along the common connection of D1, R2, and C1 and pins 1 and 2. They are all connected together as they are show in Figure L19-1.

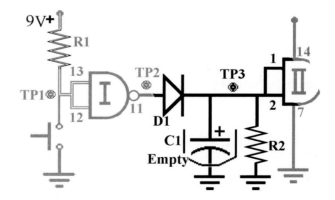

Figure L19-1

Record all of your results on the data table at the end of Section Five.

1. Before you take any measurements, allow the circuit to remain unused for at least a minute. Measure the voltage at TP 3 with the push button plunger untouched and record your results.

2. Measure the voltage at TP 3 with the push button held closed and record your results. Figure L19-2 shows that as long as the NAND gate's output at pin 11 is high, the voltage and current push through D1 and does two things. It influences the inputs of the second NAND gate and fills the capacitor.

3. Now keep the probe at TP3. Release the PB and watch the DMM. What happens here is completely different from TP1 or TP2. The voltage slowly decreases. If you are using a poor quality DMM, the voltage will drain through the probe in less than 2 seconds. Ideally, it should take over 20 seconds to drain to near 0 volts.

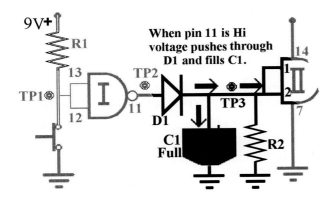

Figure L19-2

What Is Happening Here?

The illustrations shown in Figure L19-3 and Figure L19-4 show that when the PB is released the output at pin 11 goes low. The diode traps the voltage on the "right" side. The capacitor holds the voltage that influences the inputs of the second NAND gate at pins 1 and 2. Meanwhile, the voltage is draining from C1 through R2.

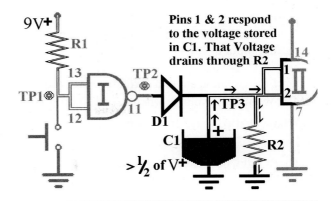

Figure L19-3

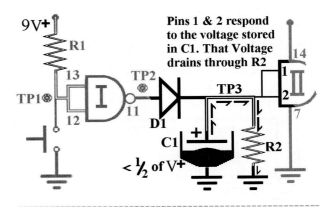

Figure L19-4

You can see the voltage draining on your DMM, too.

When a resistor and capacitor are used together like this for timing, this is referred to as an *RC circuit*.

Now consider how this affects the inputs to the second NAND gate.

- The voltage at pins 1 and 2 has just moved from 0 (low) to 9 volts (high) instantly. No problem.

- But the most important point here is that this is DIGITAL. It works in highs or lows, ons and offs.

- **The RC's** voltage is ANALOG. Inputs are designed to respond to analog inputs. The input for the second NAND gate is sliding downward, controlled by the RC circuit. Where along that sinking voltage path do the second gate's inputs at pins 1 and 2 switch from sensing the voltage as high to sensing that input as low?

Start the circuit again and carefully watch the DMM. What is the voltage when the LED turns off?

What's that you say? It looks pretty close to 4.5 volts.

That's right, but only if you use precisely 9 volts here.

Note that Figure L19-5 shows how the falling voltage would be shown on an oscilloscope. The vertical lines are used to measure time. The time units here are very large for electronics, representing half-second units.

Remember that *digital inputs* are designed to

- Sense anything above 1/2 of V+ as high.

- Sense anything below 1/2 of V+ as low.

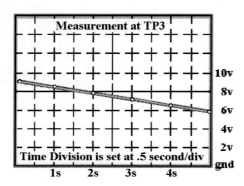

Figure L19-5

- The state of the input changes at half of the voltage supplied to pin 14.

- If the voltage powering the chip through pin 14 is 9 volts, the state of the inputs to the gate changes from high to low at 4.5 volts.

- If the supply voltage is 12 volts, the inputs to the gate will change from high to low input at 6 volts.

1. Record the "at rest" voltage of TP 3 on the data sheet at the end of Section Five.

2. Push the button's plunger and hold it closed while you check and record on the Data Sheet the voltage at TP 3.

3. How much voltage is used up by the diode (tp2@HI − tp3@HI = Vused)? Diode voltage = ____ V

4. Describe what happens to the voltage on the DMM when you release the push button.

5. How long did the LED stay on for my circuit? _____ s

6. What was the voltage on your DMM for TP 3 when the LED turned off? _____ V

7. How much time does it take for the voltage to drain to 1 volt? _____ s

8. What is the capacitor in the RC circuit being compared to? A _____

9. What is the resistor in the RC circuit compared to? The _____

10. Replace R2 with a 20-megohm resistor. How much time does the LED stay on now? _____ s

11. For R2 = 20 MΩ, what is the voltage when the LED turns off? _____ V. Is the LED going off at almost the exact point when the capacitor is half drained? Yes or No? It should be very nearly the same as before. Why would you expect this? _____

12. With R2 = 20 MΩ, how much time does it take the voltage to drain to 1 volt? _____ s How does that relate to the time it took to drain when it was the 10-megohm resistor? _____

13. Reset R2 back to 10 megohm. Replace the capacitor with the value of 10 μF. What is the time for the LED to stay lit with the capacitor 10 times larger? _____ s Was that predictable?

14. Use this as a rough formula for RC timers. It gives an *estimate* of the time it takes an RC circuit to drain from being filled to when it is near the voltage that affects the digital inputs.

R*C = T

- R is in ohms.
- C is in farads.
- T is time measured in seconds.

Here

$$C = 1 \ \mu F = 0.000001F = 1 \times 10^{-6}F$$

Also

$$R = 10 \ M\Omega = 10,000,000 \ \Omega$$
$$R \times C = T$$
$$(1 \times 10^7 \ \Omega) \times (10^{-6}F) = 10^1 s$$

See Table L19-1.

Table L19-1   Component values

	Capacitor C1	Resistor R2	Expected Time On	Real Time
1	1 µF	10 MΩ	10 S	_____
2	1 µF	20 MΩ	20 S	_____
3	10 µF	1 MΩ	_____	_____
4	10 µF	2.2 MΩ	_____	_____
5	10 µF	4.7 MΩ	_____	_____

15. Now check each RC timer in your circuit.

See if your predictions are close. They should be in the ballpark.

This is not a precise timer. RC circuits are as accurate as the components that make them. Consider what affects their accuracy.

- The resistors supplied have a tolerance of 5%.
- Aluminum electrolytic caps generally have a tolerance of 20%.

16. Look at your predictions and results. There should be an obvious pattern. Describe the pattern you see developing.

Lesson 20: Test Point 4— The Inputs Are Switches

You now get a close-up view of the output of this circuit. The LED is removed so you can get "clean" voltage measurements. Also, you start to play with the circuit's output.

Remove the LED as shown in the schematic in Figure L20-1 for measuring the voltage at TP 4. Remember to record all of your results on the data table at the end of Section Five.

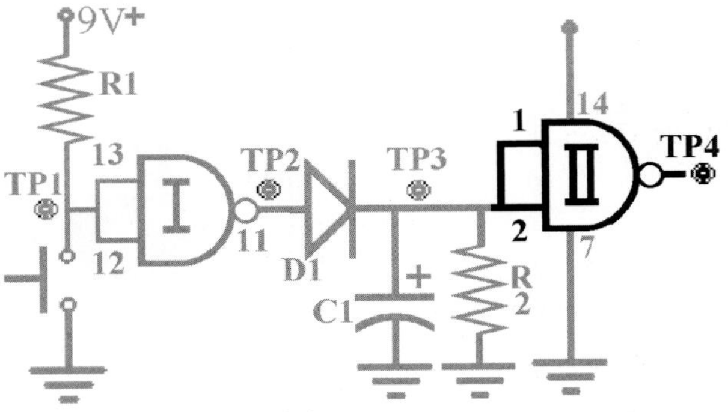

Figure L20-1

Record the voltage at TP 4 while the circuit is at rest.

Now push the plunger of the push button. It isn't necessary to continue holding it down. Record the voltage while the circuit is active.

What Is Happening Here?

Figure L20-2 shows clearly what is occurring in the circuit when it is at rest. There is no voltage stored in C1. The inputs to the second NAND gate at pins 1 and 2 are low.

When the push button's plunger is pushed, the output of the first NAND gate goes high, filling C1 and providing a voltage nearly equal to V+ to the inputs of the second NAND gate. This is shown in Figure L20-3.

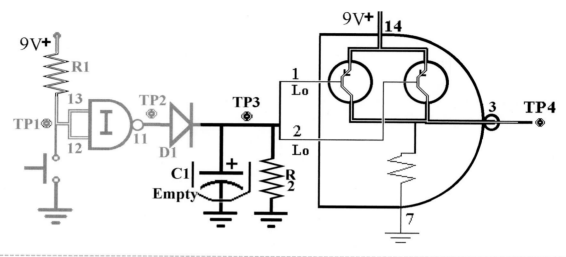

Figure L20-2

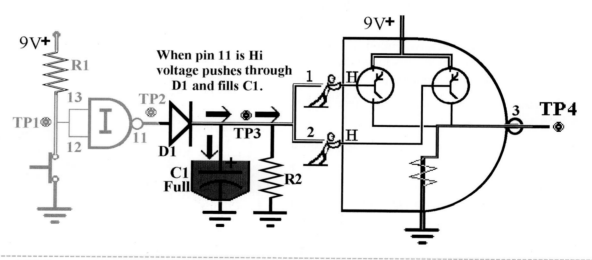

Figure L20-3

When the push button is released, as shown in Figure L20-4, the capacitor starts to drain through R2. But as long as the inputs at pins 1 and 2 sense a voltage greater than ½ of the voltage provided to the system, they continue to see this as a high input.

When the voltage stored in C1 drops below the half-way mark as shown in Figure L20-5, the inputs

now see the analog voltage as low. The internal circuit reacts appropriately. The switches reroute the voltage supplied to the chip to the output of the NAND gate. Pin 3 goes high.

Now put the LED back into the circuit, as shown in Figure L20-6.

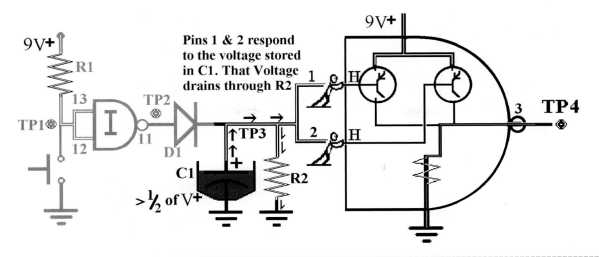

Pins 1 & 2 respond to the voltage stored in C1. That Voltage drains through R2

9V+

R1

13

TP1

12

TP2

11

D1

C1

> ½ of V+

TP3

R2

1   H

2   H

9V+

3   TP4

Figure L20-4

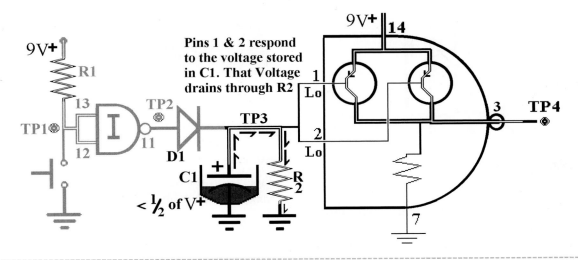

Pins 1 & 2 respond to the voltage stored in C1. That Voltage drains through R2

9V+

R1

13

TP1

12

I

TP2

11

D1

C1

< ½ of V+

TP3

R 2

9V+   14

1
Lo

2
Lo

3   TP4

7

Figure L20-5

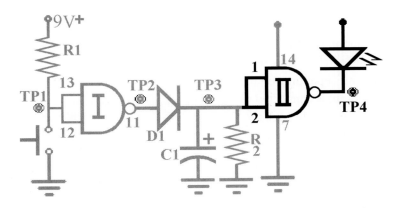

9V+

R1

13

TP1

12

I

TP2

11

D1

TP3

C1

R 2

1

2

II

14

7

TP4

Figure L20-6

There is some further testing to be done, but first, a puzzling question in two parts.

Part 1: Why does the LED stay off when the output from the second NAND gate at pin 3 goes high?

In an effort to answer this question, try this.

1. Stand up.

2. Put your palms together in front of your chest.

3. Push them together with equal force.

Why don't they move? Because each hand has equal force pushing in opposite directions. One force cancels the other. There is no movement. This concept is also shown in Figure L20-7.

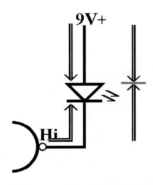

9V+

Hi

Figure L20-7

The LED stays off when the output at pin 3 provides an equal force to V+. It is like having both legs of the LED connected to voltage at the same time. Nothing is going to happen.

Part II. Remember, there were two parts to the question.

How does a "low" output of 0 volts turn the LED on?

Look at the schematic diagram in Figure L20-4 closely. The inputs act only as switches. They reroute the output to voltage or ground depending on the conditions. Here, when the output is connected to ground, the voltage moves from V+, through the LED, through the IC, to pin 7 which is connected directly to ground.

Think about the beauty of it.

- When a digital output is HI, it can be used as a voltage source.

- And when the output of a digital system is low, it can be used as ground. Figure L20-8 gives pause for thought.

Figure L20-8

Exercise: TP 4—The Inputs Are Switches

Table L20-1 is a detailed outline of the system at rest. Make a detailed outline for the active system.

Table L20-1   Outline of system at rest

Input	At rest, the inputs to the first NAND gate at pins 12 and 13 are tied to V+ through R1.
Processor	1. Because the inputs to the first NAND gate are both high, the output is low. 2. The capacitor has little or no voltage because it has drained through R2. 3. The inputs to the second NAND gate are both low. They are tied to ground through R2. 4. Because the inputs to the second gate are both low, the output from the second gate is high. The second gate acts as a source for voltage.
Output	Because there is no voltage difference between V+ and the high output of the second NAND gate, there is no pressure to push the current. The LED remains off.

---

1. Make a detailed outline for the ACTIVE system.

Input

Processor

Output

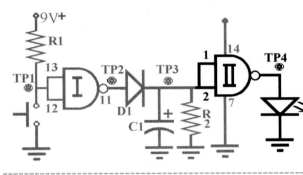

Figure L20-9

2. Connect the LED as shown in Figure L20-9.

   a. Describe what the circuit does now when it is at rest.

   b. Describe what the circuit does now when it is active.

   c. Explain what is happening. Keep in mind the fact that pin 3 is the output of the second NAND gate.

Section Five Data Sheet

Record all of your results from Section Five in Table L20-2 data table. There appears to be a huge amount of information. This single table can be used for your convenient reference and review.

Table L20-2 Data table

	System at Rest		System Active	
	PB unpushed or 1 minute	PB pushed	Immediately upon release
TP 1			
TP 2			
TP 3*			
TP 4 A (no LED)			
TP 4 B (LED In)			

*What was the voltage to the inputs of the second NAND gate when they sensed a change from a high input to low inputs and the LED turned off?

Analog Switches for Digital Circuits

Is there really gold at the end of a rainbow? That's *only wishing for power*. But the knowledge of how to use voltage dividers? ***THAT is POWER!*** There is real power in this knowledge! Knowing this gives you the power to control. Electronics is about control.

Lesson 21: Understanding Voltage Dividers

You know that voltage is used as it passes through resistor loads. The higher the value of the resistor, the greater the amount of voltage used. You already know, as well, that all the voltage is used from V+ to ground. So far, so good. Here, we use two resistors to give us any voltage we want at the midpoint. We divide the voltage. There is even a simple math formula to predict the outcome.

Here we will apply our knowledge to build switches that control the circuit. Some have moving parts, but others don't. The projects you will be building can use many types of switches. Here are some examples.

Figure L21-1 shows a simple motion detector. You can make many physical switches that work like a push button.

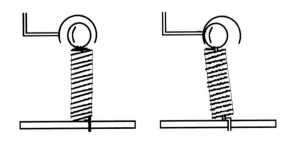

Figure L21-1

- It can be made to be small enough and be hidden in small boxes or cans.

- It can be made sensitive enough to trigger when a person walks by on a wooden floor.

Figure L21-2 uses what is referred to as a *break beam*. The sensor acts like a dark detector. It requires a source of light to keep the light-dependent resistor at a low resistance. If the beam of light is interrupted, the resistance increases. The changing voltage level triggers the inputs.

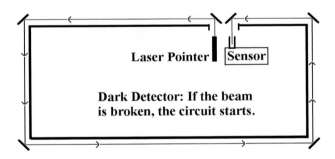

Laser Pointer **Sensor**

Dark Detector: If the beam is broken, the circuit starts.

Figure L21-2

By trading positions of the resistor and LDR, you create a light detector. Such a device can be used to alert you when a car turns into your driveway. The headlights would trigger this unit.

At what voltage do digital inputs sense the change from high to low? Ideally, it would be right at half of the voltage supplied to the system.

That means that if V+ is 9 volts, the inputs sense the high-to-low change as V+ falls below 4.5 volts.

How do we get the inputs to change from high to low at the first gate? Right now, as you can see in Figure L21-3, you have the normally open PB to connect those inputs directly to ground when you push that plunger.

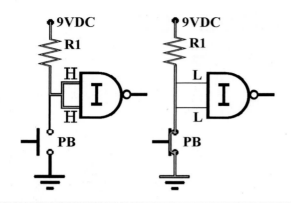

Figure L21-3

Modify the Circuit

Remember always to detach your power when making changes.

Make the three changes on your solderless breadboard that are shown in Figure L21-4.

1. Replace the PB with the trimming potentiometer.

2. Replace R1 with 39-kilo-ohm resistor.

3. Remove C1. You remove the capacitor so the circuit reacts instantly.

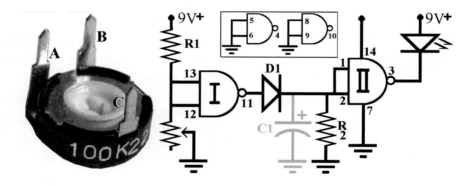

Figure L21-4

What to Expect

1. Turn the trim potentiometer to full resistance.

2. Measure resistance between A and center. (B is unconnected.)

3. It should be near 100,000 ohms.

4. Attach the power.

5. The LED should be off.

6. Adjust the trim pot until the LED goes on. It should stay on.

7. Disconnect the power; remove the trim pot.

8. Measure and note the resistance between A and center now.

Now put the trim pot back into the circuit. Turn the trim pot backward until the LED goes off. Remove it again and measure the resistance at that point. Actually, it will be much less than 39 kilo-ohms. But this is an introduction to voltage dividers. Let's keep it simple.

How It Works

We can use different resistors and variable resistors to create changing voltages similar to what we did with the night light.

Remember the night light? This is for reference only. Don't rebuild it.

Think of how the night light worked. Use Figure L21-5 as a reference.

- The NPN transistor needed positive voltage to its base to turn on.

- The potentiometer adjusted the amount of voltage shared by the 22-kilo-ohm resistor and the LDR.

- In light, the LDR had a low resistance, allowing all of the voltage to flow through to

ground. Because the base of Q1 got no voltage, the valve from C to E stayed closed.

- The resistance in the LDR increased as it got darker, providing more voltage to the base of the transistor, pushing the valve open.

- As the voltage flowed through the transistor, the LEDs turned on.

We can apply the same idea to the digital circuit inputs, as we can see in Figure L21-6.

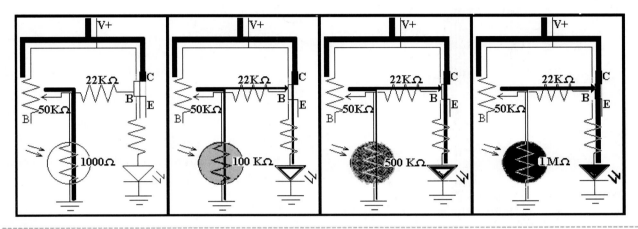

Figure L21-5

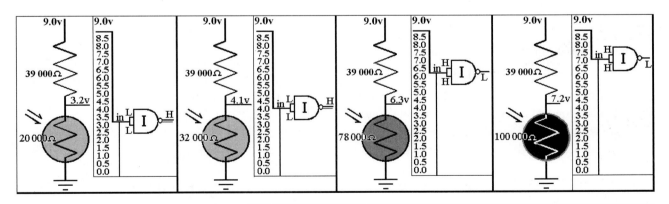

Figure L21-6

Remember: By simple definition, a circuit uses all of the voltage between the source and ground.

1. Two resistors set between voltage and ground use all of the voltage.

2. The first resistor uses some of the voltage, and the second uses the rest.

3. If you know the value of each resistor, you can figure the voltage used by each one using simple ratios. You compare the partial load to the whole load.

Look at Figure L21-7 as an example. Here R1 = R2.

```
R1 = 10 kΩ

R1 = 10 kΩ
```

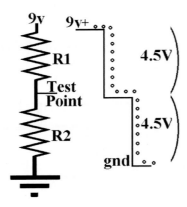

Figure L21-7

When R1 = R2, the voltage at the midpoint is exactly half of V+ because each resistor uses exactly half of the voltage.

```
V(R1/R1 + R2) = voltage used by R1

9 V (10 kΩ/10 kΩ + 10 kΩ)
```

This gives us a number of 4.5 volts used.

Why does the voltage split like this?

Simply because the resistor is a load. The larger the individual load when compared to the total load, the more voltage is used up. If there are two loads of equal value, they both use the same amount of voltage.

Build on Your Breadboard R1 = R2

Do not take apart your digital circuit.

Build the setup of two resistors on a separate spot of your SBB.

```
R1 = 10 kΩ

R2 = 10 kV
```

Measure and record the voltage at these points.

Voltage from V+ to ground_____ = total voltage.

Voltage across R1, from V+ to midpoint_____ = voltage used by R1.

Voltage across R2, from TP to ground_____ = voltage used by R2.

The voltage measured across R1 and R2 should be the same and equal to 1/2 of V+. It may be off by a few hundredths of a volt because of the following:

1. The voltmeter acts as a third load and affects the circuit.

2. The resistors have a range of accuracy of plus or minus 5%. That means a 10-kilo-ohm resistor could have a value of 9,500 ohms to 10,500 ohms.

The ideal statement for voltage at the midpoint when R1 = R2 is that the voltage is divided by half.

The load uses that portion of the voltage, in a ratio compared to the total load.

Build on Your Breadboard R1 > R2

What happens when we build a voltage divider of unequal parts?

Here is what happens when R1 is 10 times the value of R2 as laid out in Figure L21-8. Replace R2 with a 1-kilo-ohm resistor.

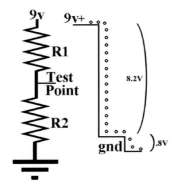

Figure L21-8

If, for example,

```
V+ = 9.0 V
R1 = 10 kΩ
R2 = 1 kΩ
```

Again, figure the voltage used like this.

$$V_{total} \left(\frac{R1}{R1 + R2} \right) = V_{used}$$

```
9 V(10 kΩ/10 kΩ + 1 kΩ) = 8.2 V
```

But remember that the important information really is the voltage remaining.

R1 uses up 10/11 of the total voltage. That is 8.2 volts to be exact. The voltage left at the midpoint should be about 1/11 of the total because

```
Total volts − used volts = remaining
                           volts

11/11 − 10/11 = 1/11

9 V − 8.2 V = 0.8 V
```

It is important to be able to predict the voltage at the midpoint.

This allows control to trigger a digital circuit with whatever switch you create.

Exercise: Understanding Voltage Dividers

Predict the voltage in a voltage divider if you have exactly 9 volts.

Don't breadboard these voltage dividers.

Use the formula. And don't forget.

```
Total − Used = Remaining.
```

1. R1 = 1 kΩ
    R2 = 10 kΩ
    Midpoint V = ___

2. R1 = 100 Ω
    R2 = 1 kΩ
    Midpoint V = ___

3. R1 = 1 kΩ
    R2 = 100 Ω
    Midpoint V = ___

4. R1 = 39 kΩ
    R2 = 100 kΩ
    Midpoint V = ___

5. R1 = 39 kΩ
    R2 = 2.2 MΩ
    Midpoint V = ___

6. R1 = 2.2 MΩ
    R2 = 100 kΩ
    Midpoint V = ___

7. R1 = 100 kΩ
    R2 = 20 MΩ
    Midpoint V = ___

Key

1. 8.2 V
2. 8.2 V
3. 0.8 V
4. 6.5 V
5. 8.8 V
6. 0.4 V
7. 9.0 V

Lesson 22: Create a Light-Sensitive Switch

Remove the last section's two-resistor voltage divider setup from the SBB, so they are no longer in the way.

The inputs to the first gate are held high via connections through R1. The circuit is at rest. No, it is not off. The circuit is off only when the power is disconnected (Table L22-1).

Table L22-1 NAND logic table

Input A	Input B	Output
High	High	Low
High	Low	High
Low	High	High
Low	Low	High

Remember: Disconnect power when you change parts on the breadboard.

Refer to Figure L22-1 as we do a quick review of how the trim pot worked as a switch.

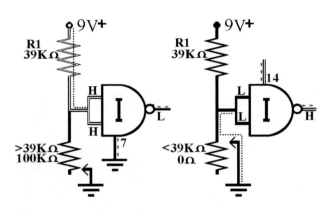

Figure L22-1

The 100-kilo-ohm trim pot replaced the normally open push button with a sliding resistance. As the trim pot changed resistance, the ratio of the voltage divider changed. Notice what happened to the output of the circuit as you adjusted the trim pot back and forth.

- As you increased the resistance, the voltage to the first NAND gate's inputs increased. It was harder for the voltage to reach ground because of the increased resistance of the trim pot.

- The inputs to the NAND gate were connected to voltage, giving a low output from the first gate.

- With the first gate inputs connected to a high, the system was at rest.

Modifying the Circuit: The Light Detector

The capacitor remains OUT (disconnected) for this lesson. Leaving it in will delay the change in output and confuse what you should see. Now try the other variable resistor, the LDR.

1. Remove the trim pot.

2. Place the LDR into your circuit as shown in Figure L22-2.

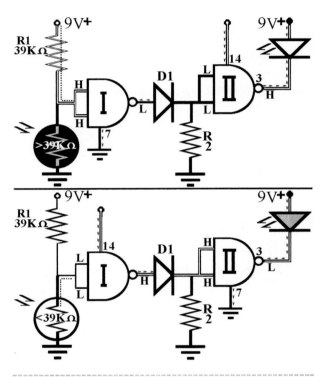

Figure L22-2

What to Expect

1. When you attach the battery, the LED should turn on instantly because this circuit is at rest in the dark.

2. Because this setup is ACTIVE in the light, you will have to place the circuit into a dark situation for it to be at REST.

3. When the resistance of the LDR goes down in the light, the inputs to the first NAND gate sense that decrease in voltage. When it drops below half of the voltage from the power supply, they sense this as a low input.

Modifying the Circuit: The Dark Detector

Reverse the positions of R1 and the LDR as shown in Figure L22-3. This simple change creates a dark detector. It is at rest in the light, and becomes active if the resistance of the LDR goes above 39 kilo-ohms.

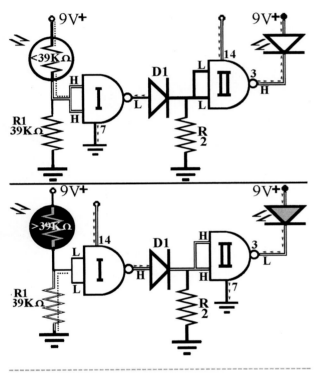

Figure L22-3

This setup needs to be in constant light to keep the circuit at rest. It will detect a break in a light source. If you put the circuit in the light and an object breaks the light source to the LDR, it will start the circuit. A common favorite is to put this switch onto a toy car. The lights turn on every time it goes under something.

Lesson 23: The Touch Switch

First, you have noticed that your skin conducts electricity, but it does have a very high resistance.

1. Set your digital multimeter to resistance.

2. Grasp a probe in each hand.

The resistance reading will constantly change, but should stay in the same range somewhere between 100,000 ohms (100 kΩ) and 1,000,000 ohms (1 MΩ).

Now change the beginning portion of the circuit to resemble the schematic shown in Figure L23-1.

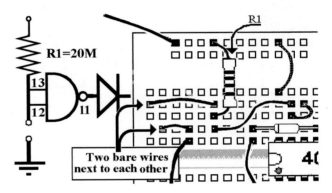

Figure L23-1

What to Expect

1. Attach your battery.

2. Touch your finger to both of the wires at the same time.

The LED turns on when your finger touches both wires that make the touch switch.

How It Works

Figure L23-2 demonstrates the effect of the finger's resistance when you become part of the circuit. The input pins 12 and 13 sense less than half of V+ when the finger touches. A voltage divider only exists when there are two resistors. So consider. Is there a voltage divider when you are not touching the contacts?

Figure the actual voltage at the inputs to the first NAND gate when the finger is touching. Assume you have a very dry finger and it has a resistance around 1 megohm.

$$V_{used} = V_{total} \times \left(\frac{R1}{R1 + R2} \right)$$

$$V_{total} - V_{used} = V_{midpoint}$$

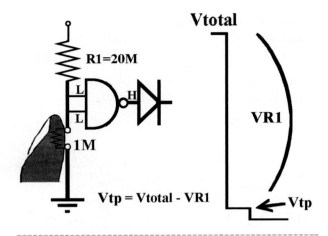

Figure L23-2

The NAND Gate Oscillator

Have you seen *The Wonderful Wizard of Oz*? Don't "ignore the man behind the curtain." I'm going to whisper some words that scare most people.

Knowledge, Design, Control

As you start to learn how to control digital inputs, you actually start to understand how some of the "whiz-bang" electronics around you actually work. Go out and buy a copy of a monthly electronics magazine. You will actually understand more than you expect. Remember, electronics is not hard, just lots of new information.

Lesson 24: Building the NAND Gate Oscillator

Here, just like the title says, you will incorporate the two unused NAND gates of the 4011 and build an extension onto your existing circuit. That extension will create a flashing output.

Add to your breadboarded circuit. Don't strip your breadboard.

Here you will get some dramatic changes by adding three basic components and changing some wiring to use the other two NAND gates.

Note what is needed in Figure L24-1 (Table L24-1).

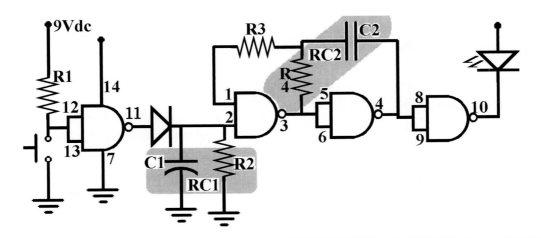

Figure L24-1

Table L24-1

Parts List

	R1	100 kΩ
	R2	10 MΩ
NEW	**R3**	**470 kΩ**
NEW	**R4**	**2.2 MΩ**
	C1	1 μF
NEW	**C2**	**.1 μF**
	D1	4148 signal diode
	LED	5 mm red
	IC1	4011 quad NAND gate
	PB	Normally open

There are only four points connected to ground now. Make sure that inputs 5/6 and 8/9 are no longer connected to GROUND.

What to Expect

Press the button to make the circuit go "active." The LED should flash once a second for about 8 seconds. It stops and automatically returns to its rest state.

If It Isn't Working: Problems and Troubleshooting

This Troubleshooting Guide will help you with the oscillator prototype on your breadboard and your finished project.

The intent of the Troubleshooting Guide is to help you locate the problem causing your circuit not to work. The hardest part of troubleshooting is finding the source of the problem. Once you find the cause of the problem, it's usually not difficult fixing it. I don't go into detail on how to fix it once you've located it. Upon closer inspection, that becomes self evident.

Something to consider. If your LED is flashing faster than 24 frames per second, it looks like it is on steadily to you. DID YOU KNOW that if the LED is blinking at 24 Hz or faster, your eyes tell you it is not blinking at all? That is why movies are shown at 26 frames per second. You are sitting in complete darkness for half of the time. You just can't notice it. Old silent movies are stuttering and jerky because they were often done at a rate just less than 24 frames per second. You can notice that.

There are usually four major problems that occur with this circuit.

1. Attach power and the LED lights up but does not blink. Start at number 1 on the table below.

2. The LED blinks as soon as you attach the battery. Start at number 1 on the table below, but pay careful attention to the first gate. Something is triggering the inputs at pins 12 and 13.

3. The LED is off until you activate the circuit. The LED turns on but does not blink. It does time off properly. Problems at RC2. Check R3, R4, and C2 connections and values. Then start at 10. If you don't get 1/2 of V+, return to start at 1.

4. The LED is off and stays off. Is your power supply connected? Start at 1. Do not just insert a fresh 4011 chip into the circuit. If a physical error blew your chip, that same error will keep on blowing chips until you fix it.

If you find a step checks out OK, then move to the next step. If not, do what is recommended.

1. Visually inspect all connections.

   - All pins on the 4011 chip should be used. If you find an open pin, something is missing.

   - A wire left in from the previous setup could still be connecting pins 1 and 2.

   - Pins 5 and 6 are connected, but the wire connecting these to ground needs to be removed.

   - Pins 8 and 9 are connected, but the wire connecting these to ground needs to be removed.

   - Make sure that none of the bare legs of the parts are touching at crossovers, creating a short circuit.

2. Look at all parts that have to be put in with polarity in mind. Positive must be toward V+ and negative toward ground.

   - Capacitors 1 μF and bigger

   - Chip

- LEDs
- Transistors (Lesson 29)
- Diodes
- Speaker (Lesson 27)

3. Check that the IC has power.

   - Note that V+ is being supplied to pin 14. Check that there is voltage being supplied from the battery to the V+ line on the board.

   - Check that there is a wire connecting pin 14 to the V+ line.

4. Note that ground has only four connections.

   Check that there is continuity from the small button on the battery clip to the ground line.

   - Pin 7
   - R2
   - C1
   - Contact to the input switch

5. Here you are looking for short circuits in your wiring. This can also be caused by sloppy soldering.

   Disconnect the power and replace the chip in the breadboard with a chip seat (empty socket).

   Get a multimeter and start checking at points noted. Infinite or over limit means that there should be absolutely no connection between the two pins with the chip removed (see Tables L24-2 and L24-3).

   Zero ohms means there is a direct connection.

Table L24-2 Measure the resistance at each leg of the chip with the black probe connected to ground.

Red Probe at	Black Probe	Expected Resistance
Pin 1	Pin 7	Infinite
Pin 2	Pin 7	Value of R2
Pin 3	Pin 7	Infinite
Pin 4	Pin 7	Infinite
Pin 5	Pin 7	Infinite
Pin 6	Pin 7	Infinite
Pin 8	Pin 7	Infinite
Pin 9	Pin 7	Infinite
Pin 10	Pin 7	Depends on the output; disconnect the output, and it should be infinite
Pin 11	Pin 7	Infinite
Pin 12	Pin 7	Infinite
Pin 13	Pin 7	Infinite
Pin 14	Pin 7	Infinite

Table L24-3

Measure the resistance from each pin to the next. The chart assumes that RC2 Oscillator is installed.

Red Probe at	Black Probe at	Expected Resistance
Pin 1	Pin 2	Infinite
Pin 2	Pin 3	Infinite
Pin 3	Pin 4	Infinite
Pin 4	Pin 5	Infinite
Pin 5	Pin 6	0 Ω
Pin 6	Pin 7	Infinite
Pin 7	Pin 8	Infinite
Pin 8	Pin 9	0 Ω
Pin 9	Pin 10	Infinite
Pin 10	Pin 11	Infinite
Pin 11	Pin 12	Infinite
Pin 12	Pin 13	0 Ω
Pin 13	Pin 14	Value of R1
Pin 14	Pin 1	Infinite

6. Replace R1 with 100 kilo-ohms (20 megohms is too sensitive and will start the circuit).

   With the battery connected, check that the voltage at pins 12 and 13 is well above half voltage when the switch is "open."

Check that the voltage at pins 12 and 13 is well below half voltage when the switch is "closed."

7. With the battery connected, check the voltage at pin 11 when the switch is "open." It should read 0.0 volts (low).

Close the switch and check that the voltage at pin 11 when the switch is "held closed." Should be V+ (HI).

- If pin 11 does not respond properly, the gate is either burnt out or pin 11 is accidentally connected to ground or somewhere else.

8. With the battery connected, check the voltage at pin 2 (RC1) when the switch is "open." It should be sinking toward 0 volts.

Then check the voltage at pin 2 (RC1) when the switch is "held closed." Should be up at full voltage.

If RC1 does not fill, check the value of R2. Also, check if diode 1 is in the right way. Then replace D1 with a power diode 1N4005. The signal diode might have burnt out. Also, check for accidental connections to ground or somewhere else.

9. When pin 2 is low, pin 3 should be high. Conversely, when pin 2 is high, pin 3 should be oscillating.

Use a multimeter to check if the oscillator is working at pin 3.

- If RC2 is set for slow pulse of 2 Hz or slower, the reading will swing from V+ to 0 volts.

- If RC2 is set for a faster frequency, the reading will stay at half of V+.

For example, if V+ is 9 volts, the meter will read 4.5 volts because it will average the voltage swings between 9 and 0 volts

10. The output at pin 3 should be directly connected to pins 5 and 6. The reading at pins 5 and 6 will be identical to the reading at pin 3.

11. The output of pins 5 and 6 is at pin 4. Check to see that the gate is working.

12. The inputs to the fourth gate, pins 8 and 9, are connected directly to pin 4. Check to see that the gate is working.

13. Is your output device working?

- Is an LED burnt out? Test them singly in a 9-volt system with a 470-ohm resistor.

- Perhaps your speaker is broken. Check continuity on the speaker wire.

- Is your transistor the correct value? Maybe it is burnt out. Table 47-2 guides you through testing of transistors.

Lesson 25: Understanding the NAND Gate Oscillator

Table L25-1 describes the present system. It is an even closer look at how the NAND gate works. The NAND gate oscillator is widely used because it can be tuned easily using an RC. Watch for the new vocabulary. Master the material now or be a slave later.

Table L25-1 Here is the system diagram as it exists now

Input	Processors	Output
Push button	RC1-Push on/timed off about 10 second	LED flashing once per second
	RC2-NAND gate RC oscillator	
	At 1 flash per second	

Recall the logic table for the NAND gate (Table L25-2).

Table L25-2 Logic table for the NAND gate

Input A (Pin 2)	Input B (Pin 1)	Output (Pin 3)
High	High	Low
High	Low	High
Low	High	High
Low	Low	High

In review, recall what happens at RC1. Gate 1 is used to start the RC1. RC1's C1 fills and drains through R2. These control the time the circuit stays on.

- Gate 1 output goes high.

- D1 traps the voltage on RC1 side.

- C1 fills.

- R2 drains the voltage.

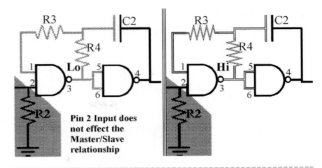

Figure L25-2

The action of an RC circuit is always the same. The only difference is the speed that the circuit fills or drains. Figure L25-1 reviews that basic action.

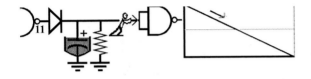

Figure L25-1

So pin 1 is a Slave to pin 3. When the system is at rest, C1 has less than half of V+. Because pin 2 is low, pin 3 gives a high output. Note the NAND gate's logic in Table L25-2. In fact, if either input is low, the output is high. This is shown plainly in Figure L25-3.

But we are interested now in RC2. RC2 is made of C2 and R4. They use gate 2 to make an oscillator. The oscillation action happens at gate 2.

Look at the setup of gate 2 shown in Figure L25-2. The high or low state of pin 3 determines the state of pin 1. Pins 1 and 3 have a special relationship. Pin 3 is the master, pin 1 is the slave.

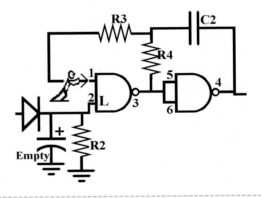

Figure L25-3

The following table shows the system at rest. There is no activity or changing voltage values happening within the circuit.

Table L25-3 shows the system at rest.

Table L25-3 System at rest

Time (Seconds)	Input A at Pin 2	Input B at Pin 1	Output at Pin 3
The system remains at rest until pin 2 changes state when RC1 gets charged.		Slaved to pin 3	As long as one input is low, the output is high.
1	Low		High
2	Low	High	High
3	Low	High	High
4	Low	High	High

But what happens when the circuit becomes active? Figure L25-4 clearly shows that when pin 2 goes high, because the capacitor is charged, pin 3 goes low.

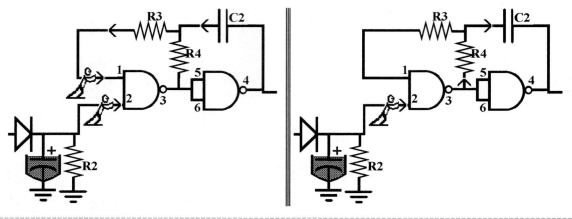

Figure L25-4

It takes a fraction of a second for the input at pin 1 to respond because C2 has to drain. Once it has drained, the voltage at pin 1 matches the output of pin 3. Wait!? Pin 3 is now low. So pin 3 makes pin 1 low, but pin 2 is high. One of the inputs is low, which makes the output at pin 3 go high. This is better than a puppy chasing its tail.

The series of actions is actually laid out very neatly in the following table.

The system starts at rest. When the system becomes active, oscillating starts as shown in Table L25-4. An animation at www.books.mcgraw-hill.com/authors/cutcher provides more detail.

Table L25-4   Series of actions

Voltage in RC1	System State	Time	Input A Pin 2	Input B Pin 1	Output Pin 3
0	Rest	0	Low		High
0	Rest	0	Low	High	High
0	Rest	0	Low	High	High
9	Active	1	High	High	Low
8.5	Active	2	High	Low	High
8.0	Active	3	High	High	Low
7.5	Active	4	High	Low	High
7.0	Active	5	High	High	Low
6.5	Active	6	High	Lo	High
6.0	Active	7	High	High	Low
5.5	Active	8	High	Low	High
5.0	Active	9	High	High	Low
4.5	Active	10	High	Low	High
4.0	Rest	11	Low	High	High
3.5	Rest	12	Low	High	High
3.0	Rest	13	Low	High	High

Animated versions of these tables at www.books .mcgraw-hill.com/authors/cutcher show the feedback loop of RC2 in action. As you can see, the feedback loop at gate 2 creates the oscillation!

Lesson 26: Controlling the Flash Rate

The NAND gate oscillator is widely used because it can be "tuned" easily using RC2. This page explains how it is done. You start to realize that there is a direct relationship between the size of the RC and the actual frequency output.

Both the values of the C2 and R4 that make the RC2 timing circuit determine the rate of oscillation. The rate of oscillation is properly called *hertz (Hz)*. Hertz is frequency per second. Another way of saying this is how many beats per second. It is a standard unit.

How It Works

First, an explanation of how the second resistor capacitor circuit (RC2) works. Then we'll play with it. This is not an exact representation of the circuit, but it will help you learn what is really happening.

Figure L26-1 shows your original RC2 setup where C2 = .1 μF and R4 = 2.2 megohms. R4 is represented by the pipeline feeding the capacitor.

Pin 2 is low. The system is at rest and stable.

The high output from pin 3 is defined by the inputs. The capacitor C2 is fully charged. There is no place for it to empty.

When RC1 gets charged, the system becomes active. You recognize that the high inputs at pin 1 and pin 2 create a low output. The low output at pin 1 allows the charge held in C2 to begin draining. It does so at a speed determined by the size of R4. Figure L26-2 shows the action of drainage from C2. As long as the voltage is above that magical half of the voltage mark, pin 1 sees its input as high.

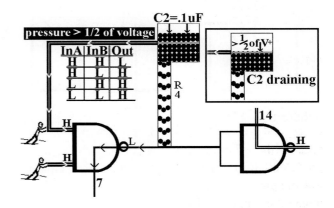

Figure L26-2

But as soon as the charge in C2 drops below a certain point, the input to pin 1 senses that input as low. HMM? Pin 2 is still high. NAND gate logic demands that the output at pin 3 become high. And C2 starts to fill (see Figure L26-3).

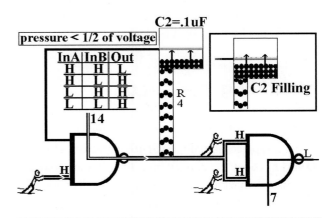

Figure L26-3

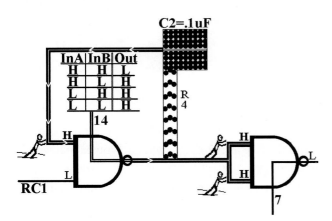

Figure L26-1

Of course, this continues until it goes above that magical marker, when the action reverses again. RC1 may be set for 10 seconds. RC2 might be set for 1 Hz. So by the time pin 2 goes low again, RC2 will have filled and drained 10 times.

RC2's rate of voltage charge/discharge is charted and shown in Figure L26-4.

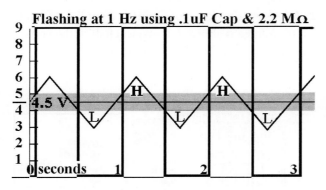

Figure L26-4

C2 fills and drains. This creates the analog input to pin 1. That sliding up and down input controls the digital high and low output shown as thick square waves.

I have made the assumption that the power supply is a convenient 9 volts. That makes the half-voltage mark 4.5 volts. Notice the grayed area around the half-voltage mark. In simplifying the explanation, I have referred to the magical point of 1/2 of V+. That's not quite true. There is a bit more of a range. If the voltage is moving upward, it has to rise above about 5 volts to be sensed as high. If the voltage is dropping, it has to drop below nearly 4 volts before it is registered by the inputs as low.

Modifying the Circuit

Make sure your battery is disconnected.

Now replace C2 with a 0.01-μF capacitor. Use the DMM to check the capacitance. Ideally, the capacitor you have is marked the same way as the disk capacitor shown in Figure L26-5. There is no standard for marking capacitors. There are several generally accepted methods. You can expect to see the marking 103Z. That refers to 10 followed by three zeros. In other words, 10,000. Disk capacitors are measured in

Figure L26-5

picofarads. That is a millionth of a microfarad. That is a thousandth of a nanofarad. 10,000 pF is 10 nF is 0.01 μF.

Or it might be marked with 0.01 or even u01. This refers to 0.01 μF. The label u01 uses the value marker as a decimal marker as well. Face it, there's not much space.

This capacitor is 10 times smaller than the one you have in the circuit right now. Capacitors this small do not have any polarity. There are no positive or negative legs. R4 is unchanged at 2.2 megohms.

Connect your power supply.

Notice the new setup shown in Figure L26-6.

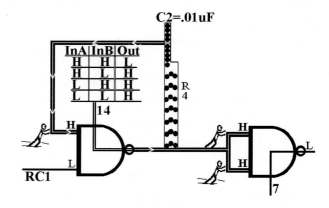

Figure L26-6

The system is at rest, but C2 is represented as a tenth the size as before. So what do you expect will happen?

The LED should flash very quickly for about 10 seconds, depending on your timing for RC1. Figure L26-7 shows the reaction by the NAND gate to the changing voltages on pin 1.

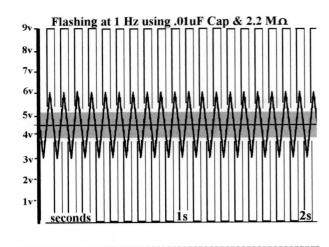

Flashing at 1 Hz using .01uF Cap & 2.2 MΩ

Figure L26-7

Ideally, it reacts exactly 10 times faster because the capacitor is 10 times smaller.

Exercise: Controlling the Flash Rate

Pull C1, the capacitor, from RC1. That way you can count without worrying about the circuit timing out at the wrong time.

On your solderless breadboard, you will change components to affect the oscillation timing of RC2. Track your results in Table L26-1.

Table L26-1   Tracking table

R4	C2	Comment	Timing Flashes in 10 s			Average
			1	2	3	
1 MΩ	0.1 μF					
2.2 MΩ	0.1 μF	Twice the resistance Expect half the rate				
4.7 MΩ	0.1 μF	Twice the resistance Expect half the rate				
10 MΩ	0.1 μF	Twice the resistance Expect half the rate				
10 MΩ	0.01 μF	Tenth the capacity Expect 10 times faster flash rate				
4.7 MΩ	0.01 μF	Half the resistance Expect twice as fast				
2.2 MΩ	0.01 μF	Half the resistance Expect twice as fast				
1 MΩ	0.01 μF	Half the resistance Expect twice as fast				

Is there a pattern when you compare it to the flashing rate using the 0.1 μF capacitor that was 10 times larger?

Lesson 27: Create a Sound Output and Annoy the Person Next to You!

You adjusted the frequency of RC2. This is a direct continuation of the previous lesson, but the LED output has a flash rate too fast to see. Did you know that people see smooth motion if related pictures are presented at 24 frames per second. That is why movies are projected onto a screen at that rate. That is also why we have to move from the LED to a speaker. When the LED is 24 frames a second or faster, it might appear to dim a little, but you won't see it flash. Why does it dim? Because it is off half the time. Don't you realize you're sitting in complete darkness in the movies for half the time, too?

Modifying Your Circuit

Don't clear your breadboard. Figure L27-1 shows the schematic that you have been using (see Table L27-1). Just replace the LED with the speaker and change values of other components stated in the parts list.

Also, if you removed C1 for the exercise in Lesson 26, put it back in.

Disconnect your power to make these changes.

Table L27-1

Parts List	
R1	20 MΩ
R2	10 MΩ
R3	470 kΩ
R4	2.2 MΩ
C1	1 μF electrolytic
C2	0.1 μF
D1	4148 signal diode
Speaker	8 Ω
IC	4011 Quad NAND gate

Going from left to right on the schematic.

1. You have a touch switch to activate the circuit.

2. The amount of time the circuit stays active is set by R2 and C1. R2 and C1 make the first resistor capacitor circuit (RC1).

3. The rate of oscillation is determined by C2 and R4 (RC2).

4. Here, the voltage from pin 10 moves from V+ (high) to ground (low) at a frequency set by RC2.

   Be cautious. Don't connect your speaker directly to the battery. Small speakers are made with very fine wire. Too much current will heat the wire, possibly enough to melt it. Such a break would render the speaker useless.

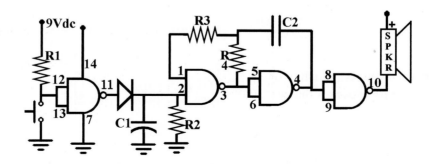

Figure L27-1

5. Speakers respond only to voltage changes. Speakers do not produce sound just because V+ is applied to them. Buzzers have a circuit inside. They create their own noise. If you put your speaker to a battery, you hear a "crackle" as you connect, and another as you disconnect. It is merely responding to changes in voltage. An excellent explanation about how a speaker works is posted at www.how-stuffworks.com/speaker1.htm.

6. Replace the LED with the 2-inch speaker. Note the polarity of the speaker.

The speaker will click slowly and very quietly. You may have to use your fingers to feel the pulse. It should pulse about 15 to 20 times in 5 seconds.

The speaker pulses each time the current is turned on, moving from low to high, and each time the current is turned off, moving from high to low. In the previous exercise you pulsed the LED at different speeds by changing the rate of oscillation in RC2.

Your exercise results should have shown this pattern (see Table L27-2).

Table L27-2   Exercise results

Resistor Values	Capacitor Values	Flashes in 10 s	Speed
10 MΩ	0.1 μF	1	Very slow
4.7 MΩ	0.1 μF	2	Double of previous
2.2 MΩ	0.1 μF	4	Doubled again
1 MΩ	0.1 μF	10	10 times faster than the 10 MΩ

Decreasing the resistance is like widening the drain. By decreasing the resistor value we increase the oscillation speed because it takes less time to fill and drain the capacitor.

Exercise: Create an Annoying Sound Output

A quick definition. Hertz: *Hertz* is a measurement of frequency, specifically defined as a measure of beats per second. For example, a system oscillating at 512 beats per second is more easily stated as 512 hertz.

As you do the following changes, note your observations in Table L27-3. Remember to detach power any time you make a change to your prototype on your breadboard.

Table L27-3   Observations

Resistor Value R4	Capacitor Value C2	Description
4.7 MΩ	0.01 μF	
2.2 MΩ	0.01 μF	
1.0 MΩ	0.01 μF	
470 kΩ	0.01 μF	
220 kΩ	0.01 μF	
100 kΩ	0.01 μF	
47 kΩ	0.01 μF	500 Hz

Now, one more change. Put your ear very close to the speaker. Listen for the quiet tone.

22 kΩ	0.01 μF	1,000 Hz

Annoying but very quiet right now because the 4011 IC does not produce very much power at the output. So the volume is not much at all. But Figure L27-2 offers hope.

You want volume?? **OK!**

But 1'st, a good oscillator deserves an oscilloscope.
The Lesson after that,
You Will Learn How To
Amplify That Output

Figure L27-2

Lesson 28: Introducing the Oscilloscope

This lesson does the following:

- Introduces one of the most important tools in electronics

- Introduces you to the concept of what any oscilloscope can do

- Shows you how to build a probe if you need to use Winscope 2.51

- Acts as an introduction to using Winscope

If you have an oscilloscope sitting on your desk at home, you are unique. If you have access to an oscilloscope, you are special.

Otherwise, you will use a freeware program called *Winscope 2.51*. You can download it from the Web site www.books.mcgraw-hill.com/authors/cutcher or from http://polly.phys.msu.su/~zeld/oscill.html. It was created by Professor Alexei Khokhlov, head of Department of Physical Chemistry of Polymers at Moscow State University.

Think of Winscope as a special "skin" for your sound card. Different skins allow for different adjustable visual effects that show on your monitor as music plays. Winscope takes this idea forward a step. The Winscope skin reacts to the internal signals as well as "microphone" or "line" inputs to your sound card.

Disclaimer

Safety measures when using oscilloscope and the oscilloscope probe.

Your scope probe is designed to be used only with your 9-volt systems. You should test it before you use it. Instructions are given in this lesson.

Being software, Winscope itself cannot damage your hardware, but it is very easy to burn out at least your sound card when trying to investigate signals of unknown amplitude and DC offset.

So, you must always be extremely careful when establishing an electrical connection between your computer and external equipment. It is a good prac-

tice to use a conventional multimeter or real oscilloscope to find out whether signal levels are acceptable for your sound card.

Regarding connecting to things besides circuits produced in this book, it is safe to connect to any audio/video equipment using standard line in jacks and cables. You may consider at least using the Scope Probe. Otherwise use an old tape recorder, amplifier, or turntable as a buffer device between your sound card and nonstandard signal source. This can save your computer in case of a poorly grounded, unstable signal source, as well as allow you to control signal level manually before it reaches the sound card.

To avoid personal injury, always follow the usual safety rules when working with electric circuits.

WINSCOPE IS SUPPLIED TO YOU AS IS, AND IN NO CASE THE AUTHOR OF WINSCOPE OR THE ASSOCIATED PROBE IS RESPONSIBLE FOR PERSONAL INJURY, HARDWARE AND/OR DATA DAMAGE, PROPERTY DAMAGE OR PROFIT LOSS ARISING FROM USE OR INABILITY TO USE THE OSCILLOSCOPE.

THE AUTHOR/DESIGNER DOES NOT GUARANTEE THE FITNESS OF WINSCOPE FOR ANY PARTICULAR PURPOSE. WINSCOPE IS NOT INTENDED FOR INDUSTRIAL OR COMMERCIAL USE.

IN GENERAL, USE WINSCOPE AND THE ASSOCIATED PROBE AT YOUR OWN RISK. ALWAYS CONSULT YOUR SOUND CARD MANUAL FOR DETAILS ON CONNECTING TO EXTERNAL DEVICES.

In the real world, we expect that as price increases, complexity and quality should increase. Winscope, however, is priceless. It still includes all of the major functions common to all oscilloscopes. Its quality is adequate for our immediate needs. One of its limitations is that it ignores stable DC voltage. But it does respond to changing voltages in the audio frequencies, 40 Hz to 15,000 Hz. Its biggest advantage is the price.

Oscilloscopes come in all shapes, sizes, and price ranges. A new scope for $500 will be barely adequate for beginners. $2,000 is not unreasonable for a good quality scope for the classroom. Reconditioned equipment is an option frequently overlooked. Such equipment that was "top of the line" for industry 20 years ago can be purchased at reasonable cost. These older tools give quality outputs comparable to equipment that would cost over $20,000 new.

Oscilloscopes are used to give instant visualization of voltage compared to time. The oscilloscope has clips to connect to the signal, and displays a graph of the voltage represented on the y axis (vertical) and time shown across the screen horizontally as the x axis. You are able to adjust both the voltage (vertical) and time (horizontal) scales.

Your DMM can measure voltage. A multimeter displays the voltage at a given moment. You can measure slow voltage changes from your DMM.

The oscilloscope specifically makes a visual picture of voltage changes that we could not otherwise see because they happen too quickly. A good scope can display voltage changes that occur in the megahertz range.

The oscilloscope display can be thought of as a graph:

- The y axis representing voltage
- The x axis representing time

If the x axis is set to 0.001 second per division, 1 millisecond, the entire screen represents events that occur in a hundredth of a second.

Figure L28-1 shows the Winscope screen.

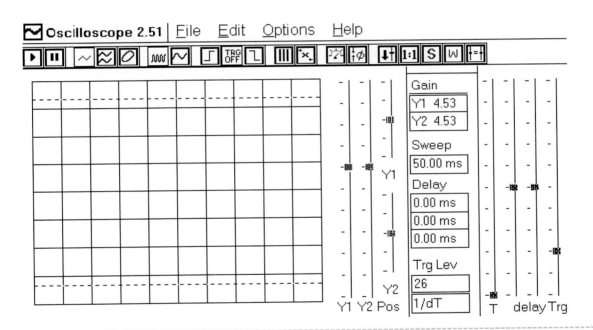

Figure L28-1

Building the Scope Probe

If you already have access to an oscilloscope, you don't need to build a scope probe. It is a piece of equipment that you need to be able to test your own circuits using Winscope 2.51 on your computer.

What the Probe Does

The probe takes any signal in your circuit, cuts out all but 1/11, using a simple voltage divider, and feeds that remaining fraction to your sound card. Your sound card feeds this signal to the Winscope software, which interprets this signal and displays it on your screen.

Keep in mind that any sound card cannot accept more than 2 volts. Anything more than 2 volts will cause damage to your sound card. So if your output is 9 volts, only .8 volts is fed to your sound card.

There are three sections to the probe as shown in Figure L28-2.

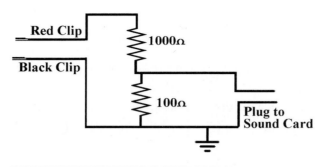

Figure L28-2

A more detailed set of photographs is available on the Web site www.books.mcgraw-hill.com/authors/cutcher.

The Connecting Clips

1. You need to have at least a 3-foot length of speaker cord. Carefully remove 1 inch of insulation from both ends.

2. Mark one of the lines on both ends to identify it as the same line on both sides. Use this line for ground.

3. Disassemble the clips and slide the covers onto the wire.

4. Twist the end of each wire strand and push this strand through the hole at the base of the clip as shown in detail in Figure L28-3.

Figure L28-3

5. Lay the insulation into the saddle and use pliers to crimp the two sides of the saddle over the insulation. This physically holds the clip to the wire.

6. Now solder the wire strand at the bottom and clip the extra wire away. You should have something similar to Figure L28-4 now.

Figure L28-4

7. Slide the covers over the back of the clips. This is done easily if the jaws are clamped open onto something large.

That should finish the clips.

The Voltage Divider

The voltage divider is the heart of the probe. It is not a regular connector, but decreases the input voltage by a factor of 1:11.

Assembling the voltage divider for the scope probe requires the following:

1. Cut the dual cord 6 inches from the end opposite the clips.

2. Strip at least 1/4 inch (0.5 cm) of insulation off four ends.

3. Mark the both sides of the ground line. This is the one connected to the black clip.

4. Figure L28-5 displays how to wrap the wire around the resistor leg before soldering. This is not critical, but it is very effective.

Figure L28-5

5. Slide skinny heat shrink tubing onto each line before you solder. The heat shrink tubing is easier to use than tape. The layout is shown here in Figure L28-6.

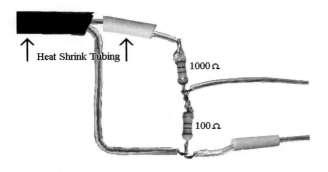

Heat Shrink Tubing

1000 Ω

100 Ω

Figure L28-6

6. Move the heat shrink tubing away from the heat of the soldering area until you are ready to shrink it into place. Test your parts placement immediately after you finish your soldering.

Use the schematic diagram in Figure L28-7 as a guide to check that your scope probe is set up properly (see Table L28-1).

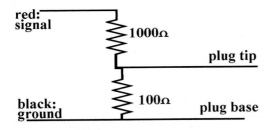

red:
signal

1000Ω

plug tip

black:
ground

100Ω

plug base

Figure L28-7

Table L28-1

Probe Values

Red clip to plug tip	1 kΩ
Plug tip to plug base	100 Ω
Ground line to plug tip	100 Ω
Signal to ground	1.1 kΩ
Ground to plug base	0 Ω

Then you can move the heat shrink tubing into place as displayed in Figure L28-8.

Figure L28-8

7. You can shrink the tubing by either using a hair dryer on its hottest setting or by caressing it with the hot solder pen.

8. Then slide the wider piece of heat shrink tubing over your voltage divider. Shrink that into place over the other heat shrink joints.

DONE.

The Jack

Connecting the plug is the same as connecting the alligator clips.

Remember to slide the heat shrink tubing onto the wire first.

1. Clamp and solder the ground wire to the long stem shown in Figure L28-9. The long stem is connected to the base, the lowest part of the plug. Now slide the heat shrink tubing over the ground line connection. Make sure there are no stray wires.

Figure L28-9

2. Lead the signal line between both of the other leads. We are going to have only one signal feed to the sound card, so both tabs will be connected. Solder that line to both tabs. There is no need to cover the last connection with heat shrink.

3. Slide the cover back down and screw it over the back.

Again, test it before you try it out. A wrong connection here can be disastrous.

You will use the scope probe in Lesson 29.

Lesson 29: Using a Transistor to Amplify the Output

Transistors are a natural choice as amplifiers. Their action and application are explained. Either transistor would work, but the PNP is used because it offers

certain advantages. You get a great response with dogs howling at 1,000 Hz. It "hertz" their ears.

So, right now you have a very quiet alarm. The output from the 4011 NAND gate provides a very small amount of power. It is enough to turn on an LED. You have already found it is definitely not enough power to give even a small speaker any volume. But it provides plenty of power to turn on a transistor.

Modifying the Circuit

Make the modification shown in Figure L29-1. Be sure to insert the PNP 3906 transistor the right way.

Why use the PNP 3906 transistor? Think . . . when the system is at rest, pin 10 is high. The two opposing voltages stopped any movement of current, and that's why the LED stayed off. Figure L29-2 shows how this action is used to our advantage. A high output keeps the voltage from moving through the transistor. The 3906 is turned off. This way, you don't drain the battery.

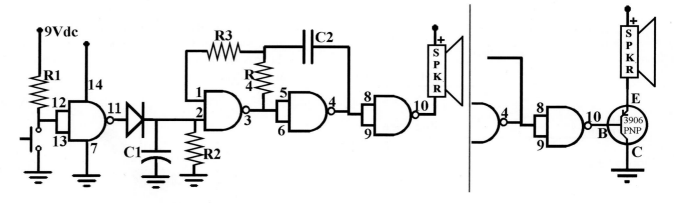

Figure L29-1

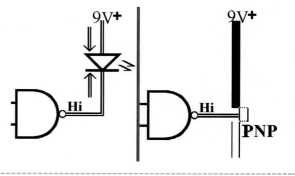

Figure L29-2

But when pin 10 goes low, the transistor turns on and the following happens.

1. This allows much more voltage and current to pass through.

2. That results in more power passing through the speaker coil.

3. The greater power produces more electro-magnetic force in the coil.

4. That produces more movement of the coil and cone, producing a louder output.

If you chose to use the NPN 3904 transistor, here's what happens. While at rest, the high output from pin 10 would keep the current flowing from voltage directly through the speaker coil to ground. This is shown in Figure L29-3.

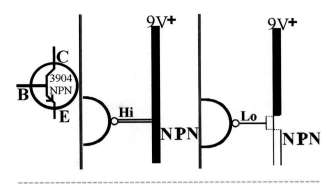

Figure L29-3

This would quickly drain the battery and annoy you.

How much better not to drain the battery quickly and go annoy some other people now.

Digital Logic Design

Wouldn't it be easier for projects just to give you a circuit board, tell you where you have to put the parts, and then solder them in? In fact, that's a great plan if your career goal is to be a solder jockey in a Third World country. However, that career opportunity is vanishing as such workers are being replaced by robots.

Lesson 30: System Design

Every system has three parts: Input, processor, output. You need to look at a range of options that you can mix and match in each section before you design your own system.

Inputs

There are four main areas (see Table L30-1) to explore in designing your own project.

Table L30-1   Four main areas

Input	RC1	RC2	Output
	Processor		
Contact Switches	Turn On/ Timed Off	No Oscillation	Low Power LEDs Music Chip
Light Detector	Time On Delay	Oscillation Rate	
Dark Detector	Touch Off		Amplified Power
Touch/ Moisture	Turn Off/ Time On		Buzzer Speaker Motors Relays

Contact Switches

Any of the contact mechanisms shown can be substituted for the push button shown in the schematic of Figure L30-1. The value of R1 should be 100 kΩ.

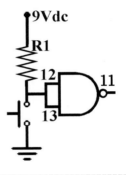

Figure L30-1

You do have the regular push button available shown in Figure L30-2. Aside from being boring, it is hard to rig these buttons to turn on with anything other than a push of the finger.

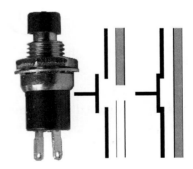

Figure L30-2

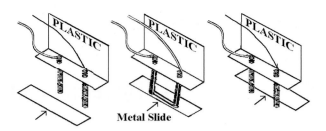

Then there is the motion detector, displayed in Figure L30-3. These can be made by balancing a weight on the end of a spring. Almost any metal weight will do, but a tapered screw is most easily attached. The best springs for this purpose are inside retractable pens. But you can't solder to spring steel. For the weight, turn the screw into the spring until it catches. For the bottom, wrap a piece of copper wire around the base. Solder the wire to the PCB. This style of switch can be made to be surprisingly sensitive.

purchase these new, they will cost upward of $4 each. Each switch has three contacts. Look closely at the photo in Figure L30-6. One is the common. Depending on your choice, this switch can act as a normally open or normally closed push button.

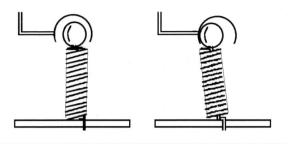

Figure L30-3

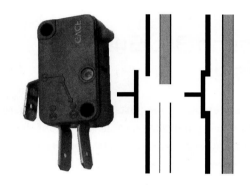

Figure L30-6

Turn the spring mechanism upside down. This pendulum setup shown in Figure L30-4 is not nearly as sensitive as the upright spring, but it uses the same concept.

A coin-activated switch is a bit trickier to make. Vending machines often use a lever on a microswitch. The coin pushes the lever down, which in turn pushes the contact switch down. A simpler device is shown in Figure L30-7.

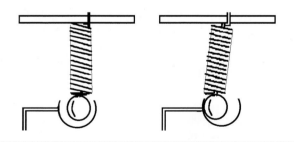

Figure L30-4

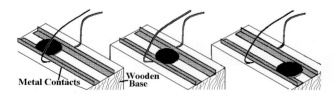

Metal Contacts   Wooden Base

Figure L30-7

Two springs can be attached to a nonmetal support. They can pass over a metal contact bar as demonstrated in Figure L30-5.

Either the spring support or the metal bar can slide. This is the perfect setup for doors or drawers.

Microswitches are exactly that. They are very small. They are readily available for free. Any broken mouse provides two of them. If you go out and try to

It works and is simple to build. Mount the two sides onto a simple wooden or plastic base.

Light-Dependent Resistor

Light Detector

The LDR is the base of this light sensitive switch. The circuit shown in Figure L30-8 will become active when it is exposed to light.

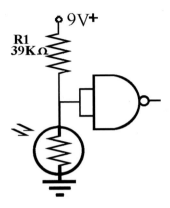

Figure L30-8

Unwanted light may turn it on rather than the event you intended. If the LDR is to be used in a generally well-lit area, it is best to use a cowling, as shown in Figure L30-9.

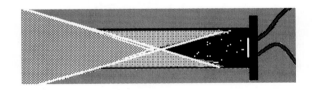

Figure L30-9

Depending on the light source, it might be necessary to use a lens to concentrate the light source onto the LDR. This is demonstrated in Figure L30-10.

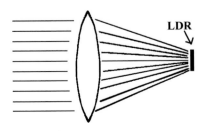

Figure L30-10

Just remember that a lens works only with a preset light source and won't successfully focus generalized light.

Dark Detector

The circuit shown in Figure L30-11 is identical to L30-8, except that the two components have traded places.

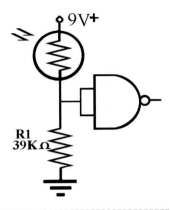

Figure L30-11

With the reversing of the LDR and 39-kilo-ohm resistor, the voltage divider is reversed as well. A cowling as shown in Figure L30-9 is even more important here. The circuit stays at rest as long as a steady light falls on the LDR. If you want to use a steady light source such as a laser pointer, the cowling guarantees the circuit will react to the breaking of that one light source.

The best source for light over a long distance is the laser pen. Using mirrors, the beam can even travel around corners. The system is shown in Figure L30-12. A laser pen can be powered with a wall

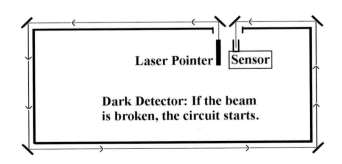

Figure L30-12

adapter matched to the same power rating as the batteries that normally power it. Every cell inside a laser pen has 1.5 volts. For example, if it has three cells, you need to find a wall adapter that provides 4.5 volts.

The beam here is shone from inside a window, and travels around the outside of the house. A speaker in both next-door neighbors' homes was set off for 10 seconds at 1,000 Hz. The system was able to be keyed off outside by the owner.

Caution: Be careful. Many laser pens claim to meet safety specification, but really can damage your eyes if you are exposed over a period of time.

Touch Switch

The schematic shown in Figure L30-13 works as a touch switch. This setup will also work with water spills. All it needs is two bare wires close together, but not touching.

Figure L30-14

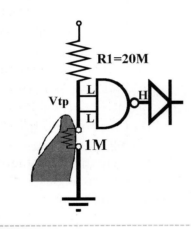

Figure L30-13

A very clean, professional-looking touch sensitive switch can be made by connecting the wires to the underside of broad-headed pins or thumb tacks pushed through black plastic. These are displayed in Figure L30-14.

Processors

Possibilities for the First Resistor/Capacitor Circuit—RC1

Figure L30-15 shows the basic RC1 setup.

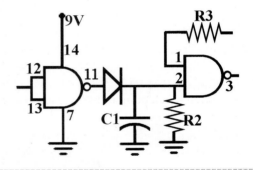

Figure L30-15

The Table L30-2 is a rough guide for timing RC1. Remember that this is only a rough guide. It is not a precise time.

Table L30-2   RC1 timing

R2	C1	Time
20 MΩ	10 μF	120
10 MΩ	10 μF	60
4.7 MΩ	10 μF	30
20 MΩ	1 μF	12
10 MΩ	1 μF	6
4.7 MΩ	1 μF	3

The schematic of Figure L30-16 shows how to manually speed up the timed off. Use pin heads for the touch switch. Your finger acts like a 1-megohm resistor. If R2 is 10 megohms, it will drain C1 ten times faster. Or you could use a PBNO to drain C1 instantly.

The schematic in Figure L30-18 is also impressive. It is a delayed time on. It can be used effectively in the light sensitive switches to slow down the triggering speed.

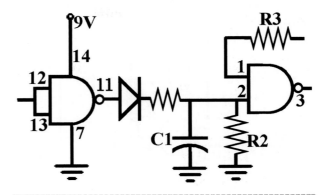

Figure L30-18

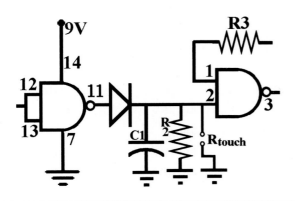

Figure L30-16

The next circuit is very similar. As you can see in Figure L30-17, the simple modification causes several changes. This alarm stays active until you turn it off. C1 keeps the inputs of the second gate high until you touch the points to drain C1. Your finger is the only drain. A hidden touch point of two pinheads or a push button (normally open) is all that you need. Use a small capacitor at C1 (0.1 μF) and when you touch the pin heads, the alarm will appear to turn off instantly (about a half second).

This modification can be used to delay the activation of the circuit. It can be used to give you time to set the circuit in a car alarm, for example, and give you time to close the door.

The value of the extra R must be *at most* a fifth the value of R2 to work because the Extra R and R2 become a voltage divider. The input at pin 2 must rise clearly above the half-voltage mark. To give more time, use a larger capacitor.

If you choose, you can remove C1 as shown in Figure L30-19. This effectively destroys RC1. When the inputs are low at pins 12 and 13, the circuit is active. The circuit immediately turns off when the inputs go high.

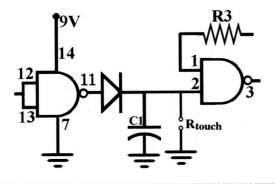

Figure L30-17

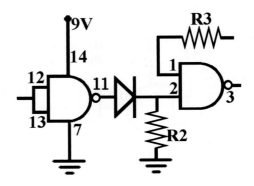

Figure L30-19

Timing and Modifications for the Second RC Circuit

There is limited potential for modifying RC2, as you can see in Figure L30-20. Either it is there and generating an oscillation at a preset frequency, or it is not there.

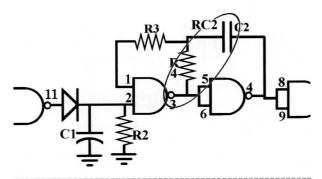

Figure L30-20

The Table L30-3 provides preset values for RC2 that produce nearly specific frequencies. It is not a precise time. You won't be able to use it as a reliable pitch pipe for tuning.

Table L30-3   Values for RC2

R4	C2	Frequency
2.2 MΩ	0.1 µF	1 Hz
2 MΩ	0.1 µF	2 Hz
470 kΩ	0.1 µF	4 Hz
220 kΩ	0.1 µF	10 Hz
100 kΩ	0.1 µF	*20 Hz
47 kΩ.1	0.1 µF	40 Hz
22 kΩ	0.1 µF	100 Hz
1 MΩ	0.01 µF	20 Hz
470 kΩ	0.01 µF	40 Hz
220 kΩ	0.01 µF	100 Hz
100 kΩ	0.01 µF	200 Hz
47 kΩ	0.01 µF	400 Hz
22 kΩ	0.01 µF	1,000 Hz
10 kΩ	0.01 µF	2,000 Hz
4.7 kΩ	0.01 µF	4,000 Hz
2.2 kΩ	0.01 µF	10,000 Hz

*The eye can't distinguish flashing from continuous motion for anything faster than 24 frames per second.

For certain applications, it is obvious that you want to remove RC2. No oscillating, please. For example, you don't want to listen to the first phrase of "happy birthday t'" (wait 2 seconds) "happy birthday t' . . . " as your circuit works through a 2-second on, 2-second off cycle. Figure L30-21 shows two details necessary to disable RC2.

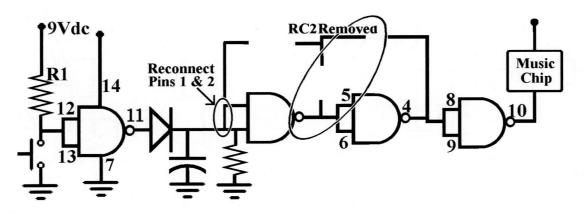

Figure L30-21

The first detail is to reconnect pins 1 and 2 together. Second, remove R3, R4, and C2. Failure to do so will lead to confusion.

Outputs

Table L30-4 describes the output by comparing oscillation needs against power requirements.

Table L30-4 Comparing oscillation needs against power requirements

	Low Power Output	High Power Output Needs a Transistor
Oscillating	Slow Flashing LEDs	Speaker for Alarm (1,000 Hz) Buzzer (slow pulse @ 1 Hz) Relay (slow pulse @ 1 Hz)
Not Oscillating	Music Chip	Car Alarm Relay (no pulse) Low-Power DC Motor

Low Power

A low-powered output is good for only low-powered applications.

LEDs

The output of a 4011 chip can power more than 10 LEDs, but not many more. Even so, there are two ways to wire these up: The right way and the wrong way. Figure L30-22 shows the right way to connect more than four LEDs.

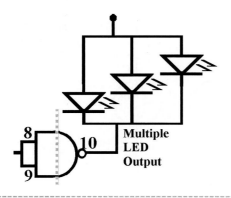

Figure L30-22

Music Chip

Carefully remove a music chip from a greeting card. Don't break any wires. Examine the music chip.

- Tape the wires to the speaker in place. Don't bend them.

- Note the circuit's connection to the "+" side of the battery.

- Remove the battery from the music chip. Remove the small stainless steel battery holder, crimped in place.

- Solder two wires, respectively, to the circuit's + and − battery connectors.

- Connect the ground side to pin 10.

- Connect the V+ side to voltage of your circuit.

- If the sound is very scratchy, place 2 or 3 LEDs in line with the music chip, as shown in Figure L30-23. The music chips run off 1.5 volts. Too much voltage can keep them from working. The LEDs use up voltage, dropping it down to where the music chip can function properly.

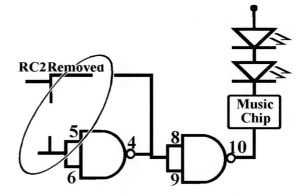

Figure L30-23

High-Powered Outputs

Buzzer

A buzzer has different needs than a speaker for output. All a buzzer needs is voltage. It produces its own signal. If you want the buzzer to turn on and off, use a slow oscillation of 1 Hz. It will "beep" once a second, on and off. A signal faster than 10 Hz will only confuse the buzzer and give muddled results at best. Figure L30-24 shows the setup.

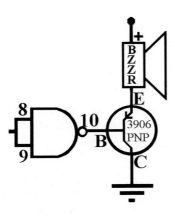

An amplified output to a buzzer should oscillate at 1 Hz. That can be created with RC2 values of R4 = 2.2 megohms and C2 = 0.1 μF (see Table L10-3).

Speakers

Speakers need a signal to be heard. If you put only voltage to a speaker, you will hear a crackle as the voltage is turned on. Nothing more. The speaker needs a signal generated by RC2. A 1,000-Hz signal generated by using a 0.01-microfard capacitor and 22-kilo-ohm resistor is a very noticeable sound. The PNP transistor shown in Figure L30-25 amplifies the strength of the signal.

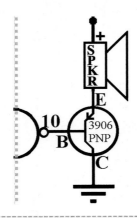

Figure L30-25

Relays

Relays allow us to use the 9-volt system to control power for another system. The on/off to the second system is connected through the relay.

Relays can be used in a variety of places. Best used in the following:

- Car alarms. A slowly pulsed relay connected to a squawker creates a sound unique from all the other car alarms we've come to ignore.

- Control the power to 120-volt circuit. This can be used for Christmas lights. The sun goes down; the lights automatically come on.

- Nonoscillating toy motor circuits, instead of direct connection to pin 10. The best results happen here when the motor uses a separate power supply and won't work off 9 volts.

This option presents easy rewiring. The on/off power to the motor is routed through the relay.

Here is a quick explanation of how a relay works. As the current flows toward ground, a magnetic field expands, creating an electromagnet that closes a switch. The *diode* shown in Figure L30-26 is *vital* because there is a close relationship between electric current and magnetism. When the electricity is turned off, the collapsing magnetic field actually pushes the current backward. The reversed diode across the relay helps to control the backward surge of voltage

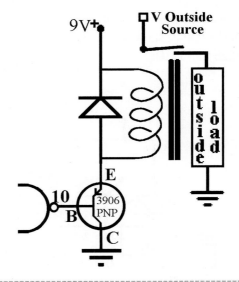

Figure L30-26

Lesson 30 – System Design

pressure and current created by the collapsing electromagnetic field. If it is not in place, the transistor will quickly burn out.

Motor

Depending on your needs, a small motor might work directly connected to a transistor as shown in Figure L30-27. The best motors for this purpose are the miniature vibrator motors made for cell phones. These can be purchased through electronics surplus suppliers found on the Internet.

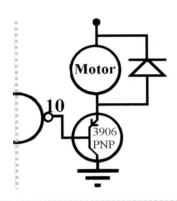

Figure L30-27

For most small motors, a relay would provide much better results.

Electric motors use electricity and also generate electricity.

Try this with the motor outside of the circuit, if you are interested.

Connect the DMM across the motor leads and spin the shaft: First one way, then the other. The motor also uses current and magnetic fields to create movement.

The reversed diode across the small motor helps to control the backward pressure of the extra voltage created by the motor. If it is not in place, the transistor would quickly burn out.

Examples

Each of the systems shown were conceived and designed by people just learning electronics.

A Pop Can Motion Detector

The weight on a spring input was "tuned" so precisely. It would start as someone walked by the table that it was on. By the time they stopped and turned around to look where the noise was coming from, it would stop. All they would see was normal looking junk on the table.

The pop had been removed via a hole in the bottom. The rim on the top had been sanded down so the lid was removed intact. A picture is shown in Figure L30-28. The related schematic is displayed in Figure L30-29 (see also Table L30-5).

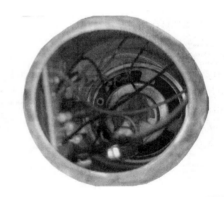

Figure L30-28

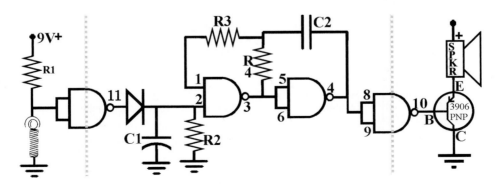

Figure L30-29

Table L30-5

Input	Processor	Output
Motion detector	RC1 = 5 s RC2 = 1,000 Hz	Speaker (amplified)

The Gassy Cow

This is definitely a young man's idea of fun. An MPG file on the Web site www.books.mcgraw-hill.com /authors/cutcher shows the real action of this fun toy. Words simply don't do it justice. A picture is shown in Figure L30-30. The related schematic is displayed in Figure L30-31 (Table L30-6).

Figure L30-30

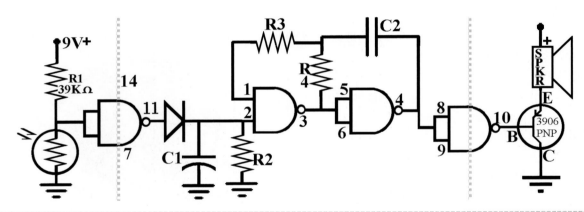

Figure L30-31

Table L30-6 Related to schematic displayed in Figure L30-31

Input	Processor	Output
Light Detector	RC1 = Instant On/Off RC2 = 80 Hz	Speaker (amplified)

Shadow Racer

An MPG file on the Web site www.books.mcgraw-hill.com/authors/cutcher shows the action of this race car. Wave your hand over the top and away it goes. It has an on/off switch; otherwise, it would be wanting to go all night. A picture is shown in Figure L30-32. The related schematic is displayed in Figure L30-33 (Table L30-7).

Figure L30-32

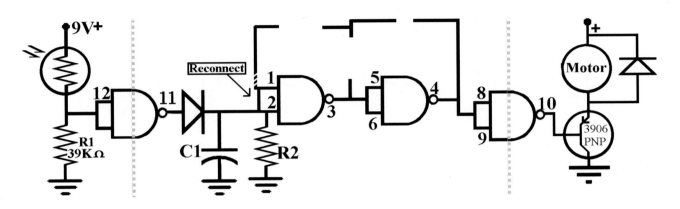

Figure L30-33

Table L30-7 Related to schematic displayed in Figure L30-33

Input	Processor	Output
Dark Detector	RC1 = 10 s RC2 = Disabled	Small Motor

Jiggle Me Teddy

This one proves that other familiar toys are no great works of genius, just great works of marketing. A picture is shown in Figure L30-34. The special motor setup for creating the jiggle is shown in Figure L30-35. The related schematic is displayed in Figure L30-36 (see Table L30-8).

Figure L30-34

Figure L30-35

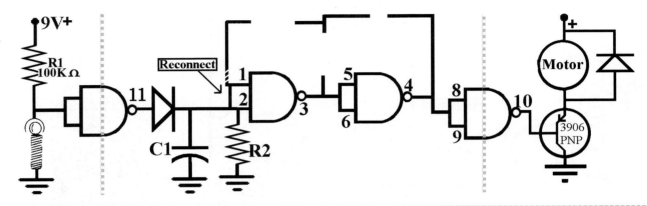

Figure L30-36

Table L30-8 Related to schematic displayed in Figure L30-36

Input	Processor	Output
Motion detector	RC1 = 10 s RC2 = Disabled	Motor with eccentric

There are two film canisters inside the Teddy Bear. The motion detector is a spring in a loop. The motor has a weight soldered onto its shaft. Both are sealed inside film canisters that keep them from getting caught up in the stuffing.

Supercheap Keyboard

This is a particularly challenging application. An MPG demonstration is given on the Web site www.books.mcgraw-hill.com/authors/cutcher.

The initial input is a touch switch, just to make the circuit active.

The second input changes the value of R4 in RC2. That changes the output frequency. RC2 is given stability by having a 20-megohm resistor connected. A picture is shown in Figure L30-37. The related schematic is displayed in Figure L30-38 (Table L30-9).

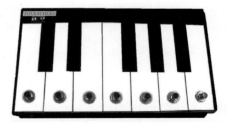

Figure L30-37

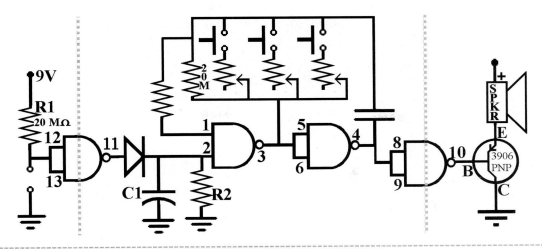

Figure L30-38

Table L30-9 Related to schematic displayed in Figure L30-38

Input	Processor	Output
Touch Switch and push buttons (RC2)	RC1 = Instant On/Off	Speaker amplified
	RC2 = Various C2 =0.01 μF R4 = Various	

Heartthrob Teddy

Even though this is a basic application, it is worth mentioning that as a child's toy, it is still a favorite. Kiss the bear on his nose and his heart throbs. A picture is shown in Figure L30-39. The related schematic is displayed in Figure L30-40 (Table L30-10).

Figure L30-39

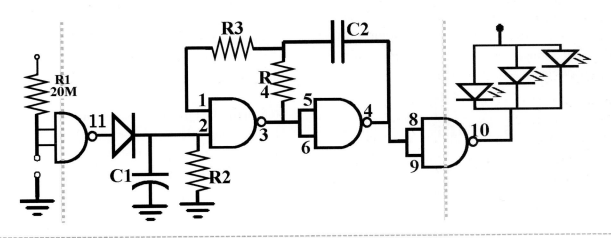

Figure L30-40

Table L30-10 Related to schematic displayed in Figure L30-40

Input	Processor	Output
Touch switch (pin heads on nose)	RC1 = 10 s RC2 = 2 flashes/second	LEDs

Lesson 31: Consider What Is Realistic

Completing this project should take no more than a few hours of applied time.

Designing the Enclosure

Use what is available around you. Consider the projects shown as examples. None of them required the students to "build" the enclosure. Each of these applications was designed with the idea of putting the circuit into something premade.

KISS

Remember the KISS principle. Keep It Simple Students! Is your initial design realistic?

Can this circuit do what you imagine it can? It is not a clock! It is *not* a radio! If you want to have two different outputs, this system is severely limited. You could have a buzzer pulsing and an LED flashing easily, but it becomes too complex to have the circuit control a music chip and a motor at the same time.

Keep It Simple Stupid Students! This is an application of your learning.

Parts, Parts, Parts

What parts are available to you?

If you purchased the kit through www.abra-electronics.com, you have what you need for the basic application, including the transistor. If you live in a larger center, there is probably an electronics components supplier in your city. Look in the yellow pages. If you don't have a supplier in town, find and order your components over the Internet. An excellent source, reasonably priced, is www.abra-electronics.com/.

The Level of Difficulty

Consider what is realistic when designing and building your project.

A simple idea applied with imagination will better impress people than a complex idea never finished.

What is a "counterproductive" design?

A motion detector in a toy car. It starts when you jiggle it, and keeps going because it jiggles itself. It is like a screen door on a submarine. It helps keep the fish out.

If you find the material fairly easy, create a simple project now so you can keep moving.

If you find the material difficult, create a simple project now so you won't get bogged down here and can keep moving through the course.

Time

Real limitations have to be balanced against time available.

There are two general things I have found in life.

Most people are always in such a rush. Do this! I gotta do that!

There is never enough time to do it right, but there is always time to do it over.

Note Regarding the LDR

Many items that use an LDR need an on/off switch. Think of your little brother or sister. Time for bed. Lights out; the doll's eyes just keep flashing.

Or you can just keep the lights on at night.

Safety

What if I want to use a relay to switch on a 120-volt AC circuit?

Your first and only answer is this. Who do you know that is comfortable working with 120 volts? Get their help.

If you don't have direct, hands-on help, use the relay to switch a smaller voltage, perhaps another 12-volt or 9-volt source to run a smaller version of your intended output.

Safety first!

You need to have a proper relay and enclosure for the 120 volts. If you want the circuit to control 120 volts, then the circuit must also be mounted properly. The PCB can be mounted properly in an enclosure.

Enclosure considerations need to meet certain standards so you don't accidentally have 120 volts to the enclosure.

Soldering considerations for 120 volts. You can have messy soldering for a 9-volt system and have it work fine, or not work. That's still safe.

Refer to Figure L31-1 to have you keep in mind that you cannot have messy soldering for 120 volts. It may spark, start a fire, or simply electrocute you.

Figure L31-1

Digital Logic Application

Here you get to solder your parts onto the printed circuit board. Yes, the same PCB is being used for all of the many different applications. The processor is essentially the same. You have developed an application by defining and designing different inputs and outputs. You also have to find a reasonable enclosure to use for your defined application.

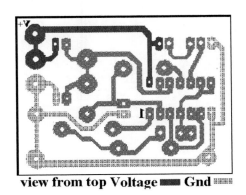

view from top Voltage ▬ Gnd ▒▒

Figure L32-2

Lesson 32: Building Your Project

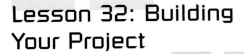

This lesson deals with parts placement. Close attention is paid to the different variations of inputs, processors, and outputs. Figure L32-1 shows the bottom view of the printed circuit board. Figure L32-2 shows the same PCB from the top view. Note that the voltage and ground lines have been displayed with different textures.

The parts placement shown in Figure L32-3 is for a standard application with low-power output. Note that the chip seat is soldered into the PCB. The 4011 can be inserted and removed as needed.

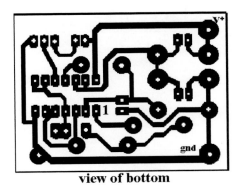

view of bottom

Figure L32-1

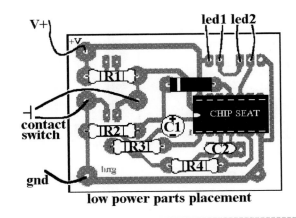

low power parts placement

Figure L32-3

Parts are shown in place for high-powered output in Figure L32-4. Again, this is a standard application for the high-power output.

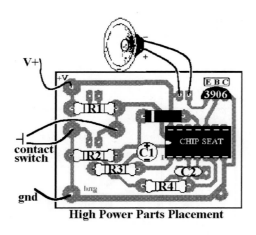

High Power Parts Placement

Figure L32-4

Inputs: Variations and Parts Placement

A close look at Figure L32-5 reveals the small difference in parts placement between the light detector and the dark detector. Use the same pads for hooking up the touch switch as shown in Figure L32-6.

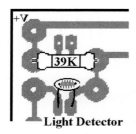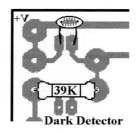

Figure L32-5

Touch Switch Inputs

out to pin heads

Figure L32-6

RCI: Variations, Timing, and Parts Placement

Remember that this is only a rough guide. It is not a precise timer (see Table L32-1).

Table L32-1   Rough guide

Resistor	Capacitor	Time Output
20 MΩ	10 μF	120 s
10 MΩ	10 μF	60 s
4.7 MΩ	10 μF	30 s
20 MΩ	1 μF	12 s
10 MΩ	1 μF	6 s
4.7 MΩ	1 μF	3 s

If you want to add the ability to adjust the timed-off setting manually, Figure L32-7 shows where to connect the wires that would lead to the contact points that you would touch.

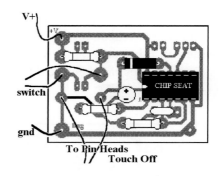

**To Pin Heads
Touch Off**

Figure L32-7

RC2 Variation and Timing

Remember that this is only a rough guide (see Table L32-2). The 4011 oscillator is not a precise timer.

Table L32-2 Rough guide*

Resistor	Capacitor	Time Output
2.2 MΩ	.1 μF	1 Hz
1 MΩ	.1 μF	2 Hz
470 kΩ	.1 μF	4 Hz
220 kΩ	.1 μF	10 Hz
100 kΩ	.1 μF	20 Hz*
47 kΩ	0.1 μF	40 Hz
22 kΩ	0.1 μF	100 Hz
1 MΩ	0.01 μF	20 Hz*
470 kΩ	0.01 μF	40 Hz
220 kΩ	0.01 μF	100 Hz
100 kΩ	0.01 μF	200 Hz
47 kΩ	0.01 μF	400 Hz
22 kΩ	0.01 μF	1,000 Hz
10 kΩ	0.01 μF	2,000 Hz
4.7 kΩ	0.01 μF	4,000 Hz
2.2 kΩ	0.01 μF	10,000 Hz

*The eye sees continuous action for anything faster than 24 Hz.

The most common option that people want, however, is to disable RC2. The most effective way to reconnect pin 1 and pin 2 is displayed in Figure L32-8.

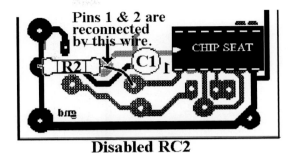

Disabled RC2

Figure L32-8

If at first your circuit does not work, then it is time to do some troubleshooting. Refer back to Lesson 24.

PART III

COUNTING SYSTEMS IN ELECTRONICS

Remember that the intention of this course is your understanding of electronics. The best way to understand electronics is by thinking of electronics as "systems." This unit focuses on introducing you to several new processors and new ways to use familiar components. The systems you can develop here are more complex than before. An entire system might have a single input or multiple settings, and there might be multiple processors combined with multiple outputs. Each component will be introduced in relation to the system (see Table PIII-1).

Table PIII-1   Parts list for Part III

Type	Description	Qty
Hardware	24-ga wire, solid core assorted	1
Semi (z)	5.1 z 1N4133	1
Semi (q)	NPN 3904	1
Semi (L)	LEDs	15
Semi (L)	Seven-seg CC 0.56"	1
Semi (d)	1N4005	5
R	100 Ω	1
R	470 Ω	15
R	1,000 Ω	1
R	22,000 Ω	1
R	47,000 Ω	1
R	100,000 Ω	15
R	220,000 Ω	1
R	470,000 Ω	1
R	1,000,000 Ω	1
R	2,200,000 Ω	1
R	10,000,000 Ω	5
R	20,000,000 Ω	4
PCB	Oscillate input PCB	1
PCB	Dual RC input 4011 PCB	1
PCB	4046 + Timed Off PCB	1
PCB	Seven-seg display PCB	1
PCB	4017 PCB	1
IC	4011	1
IC	4017	1
IC	4046	1
IC	4511	1
IC	4516	1
Hardware (sw)	PB NO	1
Hardware	Battery clip	1
Hardware	LED collars set	10
Hardware	Socket 14-pin	1
Hardware	Socket 16-pin	4
Cap	0.1 μF disk	2
Cap	1 μF radial 15 v	3
Cap	22 μF radial 15 v	1

Introducing an Analog-to-Digital Converter

Lesson 33: Introducing Possibilities—Electronics That Count

Here is a preview of some of the fun possibilities that can be designed with the knowledge you will learn in Part III. You also get a reminder regarding the care and feeding of your CMOS ICs.

Figure L33-1

The DigiDice shown in Figure L33-1 is the basic system that you will build as a prototype before you begin to design your own application.

Here a varying analog voltage input is changed into a random number generator (see Table L33-1).

Table L33-1   Random number generator

Input	Processor	Output
Push button	1. Roll down (4046 IC) controlling 2 and 3 2. Walking ring 6 LEDs (4017 IC) 3. Decimal-counting binary (4510 IC) 4. Binary-counting decimal (4511 IC) Seven-segment display	1. Fast cycling through 6 LEDs. 2. Fast cycling through numbers 0 through 9 shown on the number display. 3. Cycling slows steadily to a completely unpredictable stop. 4. Displays of both fade about 20 s after the cycling stops, and the system waits to be triggered again.

There are thousands of applications and toys that can be developed from these components. You've certainly seen some of these at the mall or casino. There are simple fortune tellers, lottery number generators, light chasers, animated signs, slot machines, and many more. You might have even spent some money on them. Ideas are explored in further depth in Lesson 43.

Your focus should be understanding how the components can relate to each other much as smaller pieces shown in Figure L33-2 relate to larger structures and models they are used to build.

Each part is a piece of the larger unit, and each unit can be a piece of the larger system.

Figure L33-2

Safety First, Last, Always

From the *CMOS Cookbook* by Don Lancaster, page 50.

"New CMOS ICs from a reliable source are almost always good and, with a little common-sense handling, practically indestructible. Possibly you will get two bad circuits per hundred from a quality distributor, maybe a few more from secondary sources unless you are buying obvious garbage. In general, the ICs are the most reliable part of your circuit and the hardest part to damage.

"If your CMOS circuit doesn't work, chances are it is your fault and not the IC's. Typical problems include

1. Forgetting to tie down inputs

2. Forgetting to debounce and sharpen input clocking signals

3. Getting the supplies connected wrong, totally unbypassed, or backward

4. Putting the ICs in upside down

5. Doing a PCB layout topside and reversing all the connections

6. Missing or loosening a pin on a socket or bending a pin over

7. Misreading a resistor (have you ever noticed the color-code similarity between a 15-ohm resistor and a 1-megohm resistor?)

And of course, causing solder splashes and hairline opens and shorts on a printed circuit board.

"The key rule is this: Always BLAME YOURSELF FIRST and the ICs LAST. Always assume that there is something incredibly wrong with your circuits when you first power them up. You'll be right almost every time. In fact, if things seem to work perfectly on the first try, this may mean that the real surprises are hiding, waiting to get you later or when it is more expensive to correct them. Anything that 'has' to be correct is usually the mistake. And what seems like 'impossible behavior' is really the poor IC trying its best to do a good job. With a little help and the right attitude you can help the ICs along."

Remember: What seems like impossible outputs really are impossible outputs.

Lesson 34: RC1—Creating the Switch

Here you are introduced to the zener diode, used inside an RC circuit. It modifies the input to the 4046 IC.

The switch for this circuit is an RC that provides voltage from 4 volts to 0 volts as the resistor/capacitor circuit drains. But doesn't my battery provide 9 volts? How do I get 4 volts? For this task, you will use a zener diode. Three common types of diodes are displayed in Figure L34-1 (Table L34-1).

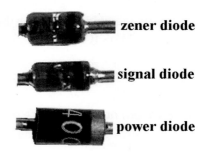

zener diode

signal diode

power diode

Figure L34-1

Table L34-1   Common types of diodes

Zener diode (bottom)	One-way valve used as counterflow lane during rush hour. It is for light traffic.
Signal diode (middle)	One-way valve to handle light traffic (smooth 1-lane unpaved road).
Power diode (top)	One-way valve able to handle large traffic demands (4-lane highway). Marked as 1N400#

A zener diode allows for two-way traffic, but only if there is enough voltage pressing backward. Here's what that means. When voltage is applied to the positive side (anode) of the zener diode, all the voltage passes through. This is shown in the left side of Figure L34-2.

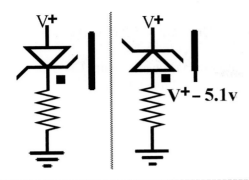

$V^+ - 5.1v$

Figure L34-2

But on the right side of Figure L34-2 the voltage is applied to the negative side (cathode) of the zener. In that situation only a predetermined amount is blocked. That is called the zener diode's *breakdown* voltage. Simply put, that is when the diode's properties break down. The zener diode we are using has a breakdown voltage rated at 5.1 volts.

It is important to note that zener diodes are labeled the same way as other diodes, but they must be put in "backward." So even though the black line still refers to the diode's cathode, as shown in Figure L34-3, a zener cathode is pointed toward the positive. Note the extra squiggle on the cathode bar that identifies the diode as a zener.

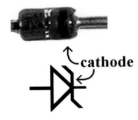

cathode

Figure L34-3

Zener diodes are available in ranges of 2 to 20 volts.

It is very important to note that your power supply must be 7 volts or more. If you use a power diode as a "protection" device, it will drop voltage by nearly 1 volt. And 7 volts minus 1 volt leaves 6 useful volts. Now the zener blocks 5 volts. Thus, 6 minus 5 volts

leaves 1 volt; 1 volt left to play with. You can't do much with 1 volt. This is particularly a concern if you are using a 9-volt battery as your power source.

Exercise: RCI—Creating the Switch

1. You are going to measure the breakdown voltage of your zener diode. It is rated at 5.1 volts.

    Figure L34-4 has two different setups for you.

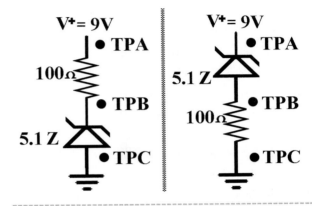

Figure L34-4

The voltage used by any component in a circuit does not change because of its position in the circuit (see Table L34-2).

Table L34-2   Two different setups

For the Setup on the Left	For the Setup on the Right
Total V from TPA to TPC = _____	Total V from TPA to TPC = _____
V used from TPA to TPB = _____	V used from TPA to TPB = _____
Volts used across zener = _____	Volts used across zener = _____
How accurate is the rating? _____	How accurate is the rating? _____

2. Now check out the actual effects of the zener diode (see Table L34-3). Figure L34-5 shows two circuits. Breadboard each separately and do the measurements. Note the LED in each.

Table L34-3   Two different setups

For the Setup on the Left	For the Setup on the Right
TPA to TPD (voltage to ground)	TPA to TPD
TPA to TPB	TPA to TPB
TPB to TPC	How much voltage is available to the LED at this point?
How much voltage is available to the LED at this point?	TPB to TPC
TPC to TPD	TPC to TPD
Describe the brightness of the LED compared to the next setup.	Describe the brightness of the LED. _____
_____	Why is the LED different brightness than the previous setup?

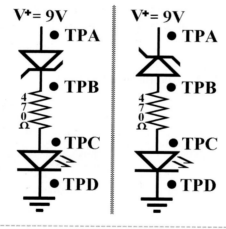

Figure L34-5

3. Define "breakdown" voltage.

_____

4. If a zener diode has a breakdown voltage of 7.9 volts, how much voltage remains after the diode when V+ is 12 volts? _____

Breadboard This Circuit

Build the circuit shown in Figure L34-6 on your breadboard (see Table L34-4).

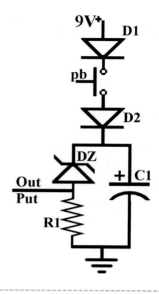

Figure L34-6

Table L34-4

Parts List
PB Normally open push button
D1, D2 Power diodes
R1 10 MΩ
DZ 5.1 Zener diode
C1 1 μF electrolytic

This is the switch that you will use for the larger system. But first, do some initial voltage testing at the output. Check for a few critical items.

1. What is the peak voltage?

2. How long does it take to drain to 0 volts?

3. How sensitive is it?

   a. Is any voltage fed through if you touch across the bottom of the push button?

   b. Substitute an LDR into the spot for the push button. Anything?

4. Change the values for R1 and C1. Does the RC act the way you would expect?

5. Take the zener diode out. Replace it with a wire. Do the measurements again. What is the biggest effect that the zener has on this switch?

Lesson 35: Introducing the 4046 Voltage-Controlled Oscillator

The 4046 chip takes the analog voltage from the RC switch and changes it directly into digital "clock" signals (Table L35-1). It is the "analog-to-digital" converter. How does it count? Digitally! $0 - 1 - 0 - 1 - 0 - 1 - 0$

Table L35-1 System diagram of our application of the 4046 IC

Input Pin 9 from RC1	Processor VCO	Output Rolldown Clock Signal
The rate of voltage drain is controlled by the value of R1 and C1 (RC1). The voltage from RC1 is moving from about 4 volts to ground.	The VCO in the 4046 compares the input voltage at pin 9 to the supply voltage at pin 16. There is a direct relation between the input voltage and the speed of oscillation at pin 4, the output. The higher the input voltage, the faster the oscillation output. The maximum oscillation is controlled by R2 and C2 (RC2).	The length of the rolldown is controlled by RC1. The output starts quickly but "rollsdown" to a complete stop. **Definition: A clock signal** is a rise from 0 v to V+ in less than 5 uS. (1 u = 1 micro = 0.000001)

The voltage input at pin 9 determines the frequency output at pin 4. Breadboard this circuit (see Table L35-2).

Pay attention to the schematic in Figure L35-1. Only three pins are connected on the top side of the IC. Don't accidentally build on top of the 4046.

Table L35-2

Parts List

R1	10 MΩ
R2	220 kΩ
C1	1 μF electrolytic cap
C2	0.1 μF Film or disk cap
D1	1N400# Power diode
D2	1N400# Power diode
DZ	1N1N4733 5.1 Zener diode
PB	Normally open PB
LED	5 mm round
IC1	4046 CMOS

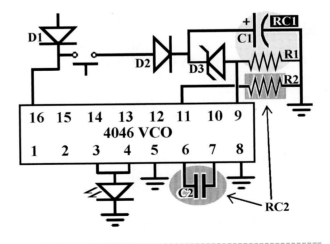

Figure L35-1

Also, note the space restriction that is being imposed. You should squish this initial circuit into the first 15 rows of the breadboard as noted in Figure L35-2.

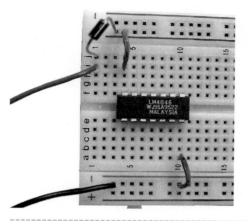

Figure L35-2

What to Expect

The circuit should work like this. You push and release the plunger. The LED turns on. As you wait, you notice that the LED appears to be flashing, but so fast, you're not sure. As you patiently watch, the flashing becomes very obvious as it slows down. It should slow down to a crawl before it stops. It could stop with the LED on or off. But it will stop. To see an animation of what to expect when the circuit is working, go to www.books.mcgraw-hill.com/authors/cutcher.

1. We are using only the voltage-controlled oscillator portion of the 4046.

2. Breadboard this circuit and briefly test that it is working.

3. Be sure to scrimp on the space as shown in Figure L35-2. This circuit should occupy only 15 rows of the SBB.

4. All components should be laid out north-south or east-west.

5. Wiring should not cross over the top of major components.

6. Wire length should be kept short enough so that your little finger can't fit underneath.

In this schematic, the wire connecting R2 to pin 11 jumps across the wire connected to pin 9.

4046 Data Sheet

Now is the time to look at the 4046 data sheet. Even if your circuit is not immediately successful, work through the *data sheet first*.

The diagram in Figure L35-3 is relevant to all ICs. Like any manufactured product, each chip is marked with all the important information. Who would have thought?

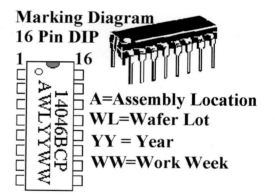

**Marking Diagram
16 Pin DIP**

1   16

14046BCP
AWLYYWW

A=Assembly Location
WL=Wafer Lot
YY = Year
WW=Work Week

Figure L35-3

The system diagram and pin assignment are both shown in Figure L35-4. As with any data sheet, these are specific to the 4046 voltage-controlled oscillator and phase comparator IC.

Lesson 35 — Introducing the 4046

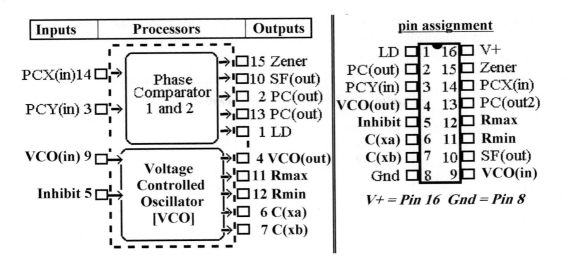

Inputs	Processors	Outputs

PCX(in)14 → Phase Comparator 1 and 2 → ☐15 Zener
PCY(in) 3 → → ☐10 SF(out)
→ ☐ 2 PC(out)
→ ☐13 PC(out)
→ ☐ 1 LD

VCO(in) 9 → Voltage Controlled Oscillator [VCO] → ☐ 4 VCO(out)
Inhibit 5 → → ☐11 Rmax
→ ☐12 Rmin
→ ☐ 6 C(xa)
→ ☐ 7 C(xb)

pin assignment

LD	1	16	V+
PC(out)	2	15	Zener
PCY(in)	3	14	PCX(in)
VCO(out)	4	13	PC(out2)
Inhibit	5	12	Rmax
C(xa)	6	11	Rmin
C(xb)	7	10	SF(out)
Gnd	8	9	VCO(in)

V+ = Pin 16   Gnd = Pin 8

Figure L35-4

1. The 4046 IC is a dual purpose chip. It has two major processors that can be used independently of each other. For our purposes, we will use the 4046's VCO ability to convert analog voltage input to digital frequency output.

2. As with all ICs, unused inputs must always be tied to an appropriate logic state (either V+ or gnd). Unused outputs must be left open.

3. The inhibit input at pin 5 must be set low. When the inhibit is high, it disables the VCO to minimizie power consumption in a standby mode.

4. Analog input to the VCO is at pin 9.

5. Digital output of a square wave is at pin 4.

6. The frequency is determined by three factors.

The voltage value at pin 9 (as compared to V+ and ground).

Cx at pins 6 and 7 and Rmx make the RC that sets the maximum frequency.

Cx and Rmn make another RC that sets the minimum frequency. The higher the resistance, the slower the minimum flash rate. In this circuit Rmn is empty; the resistance is infinite. The minimum flash rate is 0 (full stop) (see Table L35-3).

Table L35-3   Minimum flash rate

Input	Processor	Output
Voltage fed to pin 9	Input voltage at pin 9	The VCO produces a square wave at pin 4 [VCO$_{Out}$]
	[VCO$_{in}$] is compared to V+ and ground.	
	Maximum frequency output is determined by Cx and Rmx.	
	Minimum frequency output is determined by Cx and Rmn.	

An example of the expected output on an oscilloscope is provided online at www.books.mcgraw-hill.com/authors/cutcher. Examples are available for both real scopes and Winscope.

Troubleshooting PCCP

These are always the first four steps of troubleshooting:

1. **Power.** Check your voltage and ground connections. Use the DMM to check the power and ground to the chip. This inspection includes checking the power supply voltage, proper connections, and broken power connectors.

2. **Crossovers**. Visually inspect that there isn't any inappropriate touching in your circuit. Don't get your wires crossed. We're talking electronics here.

3. **Connections**. Check your connections against the schematic. For example, the schematic shows there are *only* three connections to the chip from pins 9 to 16. There are only five on the bottom, between pins 1 and 8. Did you miss one or get an extra one? Are they in the right place?

4. **Polarity**. What could affect the circuit if it were backward? Let's think about that. Hmm? Diodes, or the output LED, C1 is electrolytic, so it is polar. What about the chip? Think it would work if it were popped in backward? Don't try it out to see.

Ninety-five percent of all problems will be found if you work through the PCCP method. Predict what the output should be. Use your DMM to double check the results that you expect.

Exercises: Introducing the 4046 VCO

1. Here's a question to think about. Look carefully at the system diagram for the 4046. Why is pin 3 connected to pin 4? _____

2. In digital logic, there are two states. They are not Hawaii or Alaska. What are the two logic states? _____ and _____

3. Why must unused inputs be conditioned high or low? _____

4. RC1 determines the length of the rolldown.

   R1 = ___

   C1 = ___

   RC1 timing _____

   Replace C1 with a value 10 times larger.

5. What is the voltage being provided to your circuit? Use the DMM and measure this at pin 16. _____ volts.

6. Remove the LED and set the DMM red probe directly to pin 4, black to ground.

   Start the rolldown. Record the reading of V+ pin 4 at the beginning of the rolldown. _____ V

   As the rolldown begins, the voltage reading at pin 4 should be 1/2 of the voltage at pin 16.

7. An oscilloscope (Winscope) is the best tool to help understand this. With the LED removed, attach the scope probe's red clip to the output at pins 3 and 4. The black clip goes to ground. The plug goes into the "Line In" connection of the sound card. The settings are shown in Figure L35-5.

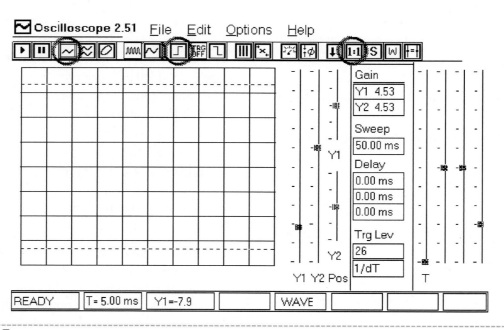

Figure L35-5

Figure L35-6 shows why the DMM reads half of the V+. The output is V+ half the time and 0 volts for the other half. The signal is switching quickly enough that the DMM averages the voltage and reads it as half of V+.

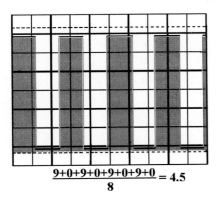

$$\frac{9+0+9+0+9+0+9+0}{8} = 4.5$$

Figure L35-6

8. A clock signal is defined as _____

9. A square wave is made by a clock signal occurring at a steady frequency. Draw a representation of a square wave in the oscilloscope face presented in Figure L35-7.

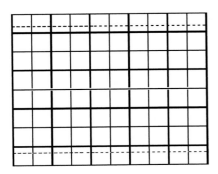

Figure L35-7

10. Modify your breadboard slightly to represent the changes shown in Figure L35-8. This allows you to adjust the voltage using a voltage divider. If the voltage is kept stable at the input to pin 9, what happens to the output at pin 4? _____

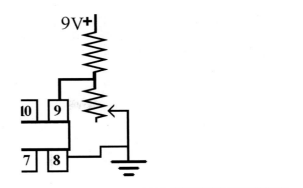

Figure L35-8

11. Consider the schematic presented in Figure L35-9.

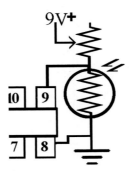

Figure L35-9

Input	Processor	Output
Amount of light	VCO	Frequency changes

12. RC2 controls the maximum frequency. It is made of Rmx and Cx, noted in the schematic as R2 and C2. Predict what would happen if the resistance were increased by 10 times.

Frequency would be

a. 10 × C faster

b. 10 × slower

Try it out. Change R2 from 220 kilo-ohms to 2.2 megohms. Did it make the frequency faster or slower? Did you predict correctly? Think about it. Did the increased resistance speed up or slow down the drain of C2? Reset R2 back to 220 kilo-ohms.

13. a. On the 4046 data sheet, pin 12 controls the minimum frequency, Rmn. Right now pin 12 is empty. No resistor means "infinite" resistance. Air is a pretty good resistor. Place a 10-M resistor from pin 12 to ground. Start the rolldown. What is the result?

b. Predict, if there is less resistance, that the minimum frequency is going to increase (faster) or decrease (closer, to no flashing when it stops?

The 4017 Walking Ring Counter

So we have a clock signal output. It rolls down to a complete stop. Now to expand from 1 LED to 11 LEDs. They don't all go on at once. It actually counts: 1, 2, 3, 4, 5, 6, 7, 8, 9, 10, 1, 2, 3, . . . and higher if you want to use the carry out capabilities.

Lesson 36: Introducing the Walking Ring 4017 Decade Counter

Here is the system diagram for the prototype application of the 4017 IC (see Table L36-1).

Table L36-1 Prototype application

Input	Processor	Output
Roll down clock signal LEDs. from the 4046 output	1. Each clock advances the high output sequentially, walking through outputs zero through nine.	1. Fast cycling. through 10
	2. When the output reaches nine, it starts counting from zero again.	2. Cycling slows steadily to a complete stop, randomly at points 1 through 10.

Be very careful not to use too much space. There are still two major circuits to fit onto your board. All of the cathode sides of the LEDs can be ganged onto the unused line on the bottom as shown in Figure L36-1 (see Table L36-2). The common resistor can connect the LED line to ground.

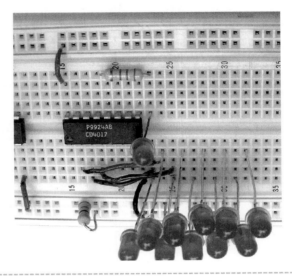

Figure L36-1

Table L36-2

Parts List	
IC2	4017 Walking ring decade counter
R3	47 kΩ
R4	10 kΩ
R5	470 Ω
R6	470 Ω
LEDs	115 mm round

Add This Circuit to Your Breadboard

The 4017 is a very simple IC. As you can see by the schematic presented in Figure L36-2, it has one major input. That is the clock signal output from the 4046 VCO.

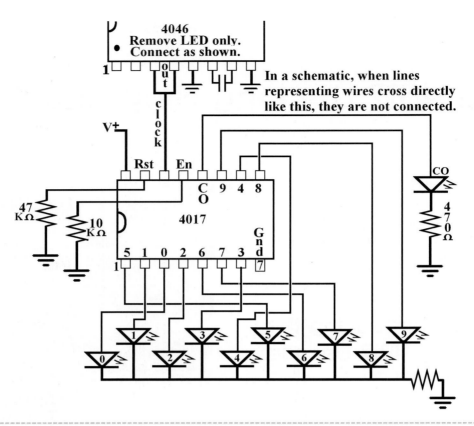

4046
Remove LED only.
Connect as shown.

In a schematic, when lines representing wires cross directly like this, they are not connected.

Figure L36-2

What to Expect

When you trigger the 4046 VCO by pushing and releasing the button, you will see the 10 LEDs lighting in sequence, appearing to have a light zipping down the series of LEDs. That is, in fact, exactly what is happening. As the VCO output slows, the zip slows as well. It will stop, randomly. The LED will be left on.

An animated schematic is available for viewing at www.books.mcgraw-hill.com/authors/cutcher. It shows the expected sequence of light with a regular clock signal input. Again, even if your circuit does not immediately work to your expectations, now is the time to look at the data sheet.

4017 Data Sheet

The system diagram and pin assignments are revealed in Figure L36-3.

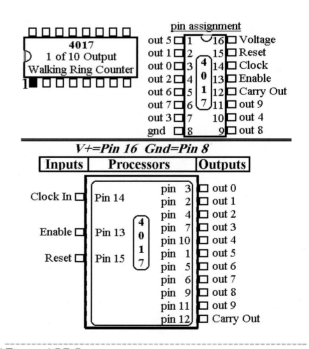

Figure L36-3

For normal operation, the enable and the reset should be at ground.

For every clock signal input, the output increases from Out(0) to Out(9) by single steps. *The counter advances one count on the positive edge (ground to voltage) of a clock signal.* Only a single output pin from Out(0) to Out(9) is high at any moment. By definition, if the output is not high, it will be low. When the count is at Out(9), the next clock input recycles output to Out(0), giving the term "walking ring" because it is walking in circles.

Carryout

When Out(9) is reached, pin 12 carryout sets high. The carryout terminal is high for counts 0 through 4 and low for counts 5 through 9.

Because the Carryout clocks once for every 10 clock inputs, the 4017 is often referred to as a "decade" or "divide by 10 counter." For example, it can be used to convert a 10,000-Hz signal input at pin 14 to a 1,000-Hz signal output at pin 12.

Making the reset high for a moment returns the counter to Out(0), setting pin 3 output to high. All other outputs drop to ground. The reset must be returned to ground to allow counting to continue. To stop the advance of the count, place a high on enable, pin 13. Returning enable to low will allow the count to continue.

As with any digital circuit, the inputs must be conditioned. That simply means they must be connected to either high or low at all times.

Of course, every input has its preferred state.

- Pin 13 enable should normally be set low, otherwise, the 4017 count will be frozen.

- Pin 15 reset should normally be set to low; otherwise, the 4017 will ignore the count. It will be stuck at 0.

The clock must be bounceless and have only one ground to positive movement for each desired count.

The 4017, like all CMOS chips, uses very little power. When it is at rest, it uses 0.002 watt of power at 5 volts.

Troubleshooting

What if the circuit is not working properly? Before you jump in, there are some bigger items to consider. Now is the time to start thinking *system*atically. Yes, you need to work through the circuit using the PCCP troubleshooting method, but the circuit just doubled in size. There are some necessary questions to ask first.

1. Do you have paper and pencil to record information down as you proceed?

2. What is working? Is there still a clock signal?

    - How can you tell? If there is no clock signal, then the problem is in the VCO.

    - If there is a clock signal and the LEDs are not working, then the problem is in the walking ring.

3. Once you have narrowed down the error to a single system area, then start the PCCP process. And even though it may sound redundant, check the power anyway.

    Here are some specific comments regarding the 4017 circuit:

    - Make sure that pin 14 has the clock signal input from the 4046 (pins 3 and 4).

    - Make sure that all of the LED cathodes (negative side) are connected together.

    - Be certain that all LEDs are connected to ground through the single 470-ohm resistor.

The most common errors are as follows:

1. Setting up the 4017 is the reversing of a single LED.

2. Not connecting pin 15 (reset) to ground. The schematic shows that it is tied to ground through a 47-kilo-ohm resistor. If pin 15 is connected to V+, output 0 remains high, and nothing else moves. If pin 15 is not connected high or low, the input will react to static electricity in the air and give unstable results. An unconnected input creates "ghosts" that are difficult to explain.

3. Not connecting pin 13 (enable) to ground. This pin is also conditioned to ground through a 10-kilo-ohm resistor. If pin 13 is connected to V+, there is no output at all the LEDs are all off. If pin 13 is not connected high or low, there will be "ghosts" created from this as well.

Exercises: The 4017 Walking Ring Counter

Reading schematics, the following drawings in Figure L36-4 show the most common styles used in schematics.

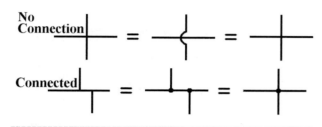

Figure L36-4

1. On the data sheet area, mention is made of the need to "condition" inputs. Explain what a conditioned input would look like.

2. What controls any integrated circuit?

3. If the inputs are not conditioned, this creates "ghosts," so called because the circuit acts in an unreliable manner and the outputs are unpredictable and unexplainable.

   • Use your DMM as an "ectoplasm-meter" to find out where these ghosts come from.

   a. Set your multimeter to AC voltage.

   b. Keep the probe wires attached to the DMM.

   c. Keep an eye on the DMM display.

   d. Quickly move your DMM toward and away from a TV or computer monitor several times. It has to be a picture tube type of display. Flat screens just don't leak enough to measure.

   e. Place the DMM right up against the monitor and hold it still.

   • What was the highest AC voltage displayed? _____AC mV

   • Do this again with DC voltage. _____DC mV

An Unconnected input is not high or low. That input reacts with the signals and static in the air around us.

Some of the signals are from

• radio and television signals

• cell phones

• the electrical wiring in the walls

4. Provide an explanation—what causes these ghosts at unconditioned inputs?

   _____

5. What is the result of having unconditioned inputs?_____

6a. There are three inputs to the 4017. What are they? (Here is some information from the 4017 data sheet.)

   a. _____ at pin _____conditioned by its connection to the 4046 output.

   b. _____ at pin _____ conditioned ____ in normal operation.

   c. _____at pin _____ conditioned ____ in normal operation.

6b. State the purpose of each input. (Here is some information from the 4017 data sheet.)

   a. _____/ purpose_____

   b. _____/ purpose_____

   c. _____/ purpose_____

7a. What is the strict definition of the clock signal?

   _____

7b. Which shape in Figure L36-5 best shows a clock signal?

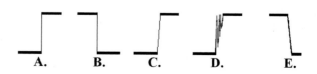

Figure L36-5

8. Set the DMM to VDC. With the red probe to 4017's pin 1, trigger the circuit. Are the results the same as the output of pins 3 and 4 of the 4046? _____

9. The output from the 4046 VCO on the oscilloscope should have been similar to the diagram shown in Figure L36-6.

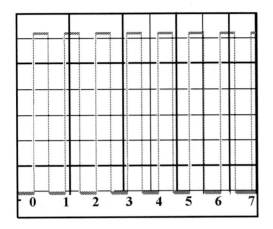

Figure L36-6

If you set up the oscilloscope just to measure at pin 1 of the 4017, what would you expect to see?

a. The same clock signal output frequency

b. Only half of the output frequency

c. High for the first five, low for the second five

d. Only one tenth of the output

10. a. Consider the carryout (pin 12). It is high for half the cycle, and low for the other half.

   When does the carryout of the 4017 change from low to high?_____

   When does the carryout of the 4017 change from high to low?_____

   b. Explain why the output of pin 12 (carryout) is so different from the other output pins 0 through 9. (Refer to the 4017 data sheet.) _____

   _____

Lesson 37: Understanding the Clock Signal and the 4017

Definition: A *clock signal* is a clean digital signal that raises from zero to full voltage instantly. It must take less than 5 microseconds. That is 0.000005 second (1 microsecond $= 10^{-6}$ second).

A clock signal is all of the following:

- Is a very clean signal

- Does not bounce or echo

- Triggers one event

The clock signal is generated by the 4046 VCO circuit and is used to trigger the 4017 circuit. Each clock signal advances the "high" output by one as shown in Figure L37-1.

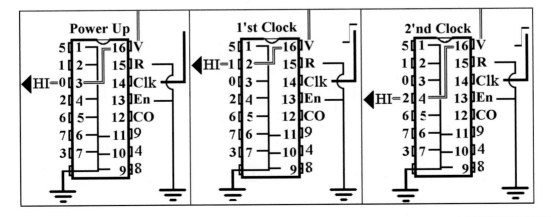

Figure L37-1

Each of the 4017's output provides a clean clock signal as it changes from low to high. When the output changes from 9 to 0, the carryout at pin 12 changes "clocks" and remains high until the count advances from Out four to Out five. Any of the outputs can be used as clock signal inputs to trigger other counters.

When reset is momentarily set to high, the count resets to 0.

When the enable is set to high, the count stops until it is returned to low.

Remember, all inputs need to be conditioned either high or low.

Refer back to the data sheet provided in Lesson 36 as you work through this lesson.

To understand the circuit better, take one of the LEDs out and put it in backward.

- Trigger the circuit.
- How many LEDs are lit at one time? _____
- There is a rolldown . . . but NOW there are two lights on all the time. One of the other nine LEDs and the reversed LED.
- The drawing in Figure L37-2 best explains what is happening.

Remember: Only one output is high at any time; the rest are low.

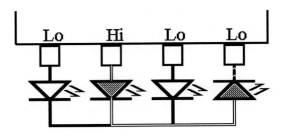

Figure L37-2

The reversed LED acts as ground. So the pin powering the LED in the count is draining through the backward LED connected to an output that is low. Return the LED to its proper position.

Exercises: Understanding the Clock Signal and the 4017

Here we will compare a PB switch to a clock signal input. This setup substitutes a single mechanical action for a clock signal. Remember that a clock signal is very fast (millionths of a second) and very clean. The LEDs advance by one per clock signal. Be sure to change back to the original setup when this exercise is done.

1. Attach a 100-kilo-ohm resistor and the PBNO to the clock input of the 4017 as shown in Figure L37-3.

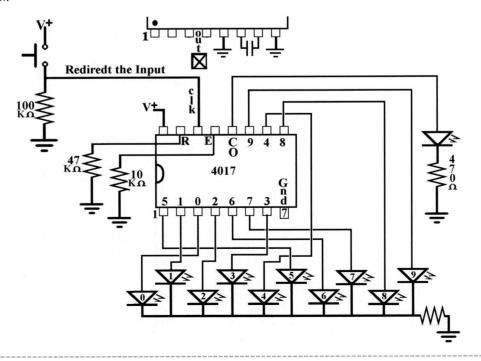

Figure L37-3

2. Remove the wire that carries the clock signal from the 4046 (pins 3 and 4) to the 4017 (pin 14). When everything is set up, attach the power.

3. What LED is lit now? LED #. _____

   - Press the plunger down in a definite movement. Don't release it.

   - What LED is lit now? LED #._____

   - Release the plunger.

   - Did the count advance when you released the plunger? That was a voltage movement from V+ to ground.

4. You expect a push button to be a fast and clean movement from gnd to V+. Is it a clock signal?____

   - You cannot use any mechanical device to provide a clock signal.

   - A clock signal must be generated electronically.

5. Push and release the plunger a few more times.

   - You can expect that the count will advance by a single step . . . sometimes by two or three steps.

   - The contacts of a physical switch like a push button do not provide a clean enough signal when a clock signal is required.

- *An unclean input has bounce* as shown in Figure L37-4. Such a bounce is too fast to be noticed on inexpensive oscilloscopes. It would be impossible to see on Winscope. But you see the results of the bounce as the 4017 counts more than one step in each push of the plunger.

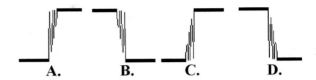

Figure L37-4

If an input says it needs a clock signal, only a clock signal will do. The CMOS 4093 is a specialized IC designed to "debounce" mechanical switches. It is nothing more than a fancy 4011 chip. It has been designed to react much more slowly to the inputs.

Lesson 38: Controlling the Count by Using Reset and Enable

Add two long wires to the breadboard as shown in Figure L38-1. These will be used as connections to V+ when needed. Leave the ends loose. Notice how pin 15 (reset) and pin 13 (enable) are each connected to ground through a resistor. That way the input is still conditioned, even if you do not have the probe connected somewhere.

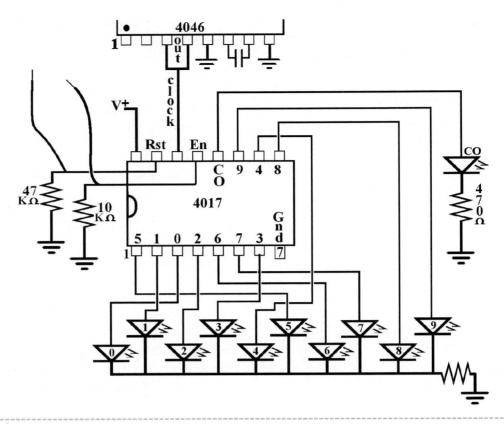

Figure L38-1

Reset

1. Trigger the 4046 to start the rolldown. All 10 LEDs should flash in sequence. Now connect the reset probe to the anode (+) side of LED number 7. What happens?

2. Choose another LED and do the same thing. What happens? What does the reset do when it is set high for a moment?

3. Wait until the rolldown stops. Touch it to V+.

4. What happens if you keep the reset high and start the rolldown?

   Take a look again at how the data sheet defines the purpose of the reset. It does exactly what it says it does. And the same is true for the enable function.

Enable

1. Pull the reset wire out while you play with the enable. Start the rolldown. Now connect the enable probe to the anode (+) side of any LED. What happens?

2. Choose another LED. What happens? What does the enable do when it is set high for a moment?

3. Wait until the rolldown stops. Touch it to V+.

4. What happens if you keep the enable high and start the rolldown?

   So what good are these? That is up to you, the user, to decide. Maybe you don't want to count zero to nine all the time. Maybe you only want it to count to six, like rolling a die. Maybe you have a Hexapod Robot. It has a preset cycle for its six legs. You can use the 4017 to trigger this system. Figure L38-2 shows a really cheap, simple, yet effective idea for applying the 4017 in a video security surveillance system.

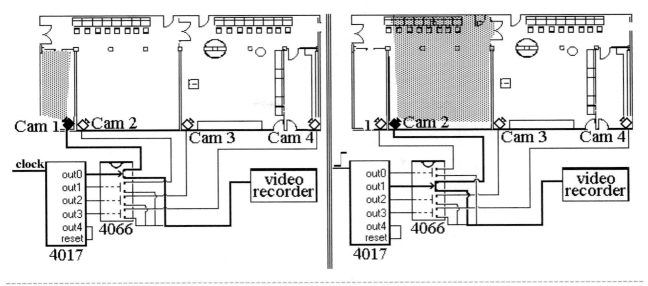

Figure L38-2

The system needs to have the "high" produced by any source to reset the system. The reset does not need a clean signal. Any momentary high will do. The "high" from pin 10 is used here. Why don't we see the LED connected to pin 10 light up? Because of the speed, the circuit reacts in microseconds. That is too fast for us to notice any response in the LED.

5. Predict what happens when reset wire is connected to pin 1. Pin 1 is related to what output?

 a. Out 0

 b. Out 2

 c. Out 4

 d. Out 5

 e. Out 10

6. Underline which LEDs light up. 0, 1, 2, 3, 4, 5, 6, 7, 8, 9

7. What is the state of the carryout for this setup?

 a. On all the time

 b. On for the first two, off for the second two

 c. Doesn't light up

8. How are the three inputs conditioned so they are always connected to either high or low?

 a. Clock_____

 b. Reset_____

 c. Enable_____

9. Look at the schematic. The "Enable" is conditioned to low through the 1-kW resistor. Start the circuit from the 4046. In the middle of this standard rolldown, connect the enable to V+. Note what happens. Leave attached for 30 seconds, and then trigger the rolldown. Disconnect the enable from V+, so it is low again. The rolldown should pick up again in the middle. Describe what the enable does.

_____

_____

10. In the following system diagram, describe the processor for the video security system shown in Figure L38-2 (see Table L38-1).

Table L38-1 System diagram

Input	Processor	Out
Video camera signal	There are three items necessary for this processor.	Videorecorder
	1. A circuit that generates clock signals.	
	2. The 4017 does what? _____	
	3. The 4066* does what? _____	

*The 4066 is the equivalent of four high-quality NPN transistors in a single package. What does it do here in the circuit?

Running a Seven-Segment Display

The seven-segment display depends on two major chips. I introduce the "slave" 4511 first, so you can understand how it is controlled. Then the 4516 "master" is introduced. You will investigate and play with most of the possibilities of each chip as they are presented. At the end of this chapter, you will have the prototype of the DigiDice and be able to explain it. Better than that, you will be able to control it. Remember, *binary*'s the word.

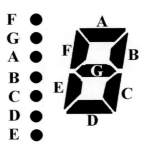

Figure L39-1

Lesson 39: Introducing the Seven-Segment Display

Here is the system diagram from the beginning of Part III (see Table L39-1).

Table L39-1   System factors

Input	Processor	Output
Push button	1. Rolldown (4046 IC) controlling 2	1. Fast cycling through six LEDs.
	2. Walking ring resets at 6 (4017 IC)	2. Fast cycling through numbers 0 through 6.
	3. Binary counting decimal (4511 IC)	3. Cycling slows steadily to a complete stop, randomly at points 1 through 6.

Remember that there is to be a number output too. In this chapter we will add the number readout. Everything revolves around creating a number output using a seven-segment LED display. This simple device is created by using individual LEDs in rectangular shape, set into a package that creates a number, as shown in Figure L39-1.

So you might have some seven-segment displays laying around. You try them out. They might work and they might not. That is because there are two major types of seven-segment displays, the CC and the CA. The type that is used here is referred to as a *common cathode* (CC). A proper schematic diagram for the CC is shown in Figure L39-2.

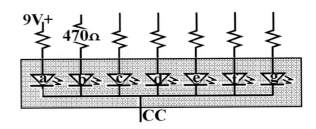

Figure L39-2

This type of seven-segment LED display is referred to as a common cathode because all of the LEDs share the same ground line. You have to add resistors on the anode (positive) side for each LED. The alternate type of seven-segment display is shown in Figure L39-3. Note that it is almost identical, but completely opposite of the CC layout.

Just as with the CC, you still have to add resistors. This is done on the cathode (negative) side for each LED in this package. Just a further note: The two

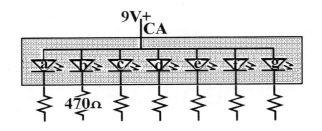

Figure L39-3

Be sure to use a 470-ohm resistor to cut the voltage to the individual LEDs as you map out the pins. The center pin of each side is a common ground connection. Only one needs to be connected to ground.

Now use a single wire to determine which pin powers which LED. Label your drawing. You will need this information to wire up your next circuit.

How would you create the number 7?

package styles are *not* interchangeable. A CC will not substitute for a CA package in any circuit because it is the same as placing an individual LED in backward. It just doesn't work.

Right now, you need to map out your seven-segment display. Set the seven-segment display into your breadboard as shown in Figure L39-4.

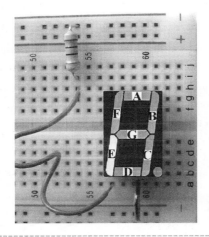

Figure L39-4

Lesson 40: Control the Seven-Segment Display Using the 4511 BCD

1. Getting numbers to present onto the seven-segment display has to be easier than rewiring all the time.

2. Several IC processors are available to change inputs into number outputs.

3. The 4511 BCD CMOS IC used here is a basic binary-counting decimal processor.

4511 Data Sheet

The system diagram and pin assignments for the 4511 BCD are shown in Figure L40-1.

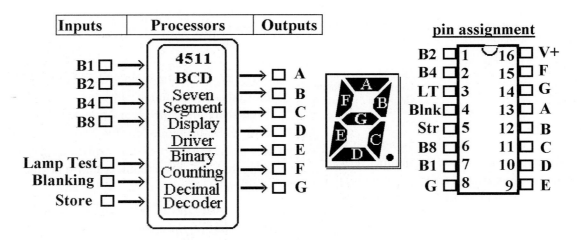

Figure L40-1

Basic Operation

The 4511 BCD accepts a binary input code and converts it to be displayed on a seven-segment, common cathode LED (all segments share the same connection to ground).

The binary code placed on B1, B2, B4, and B8 inputs is translated as a code for display on a CC seven-segment LED. For instance, if B8 is low, B4 is high, B2 is high, and B1 is low [0110], this input combination will make outputs c, d, e, f, and g high, while outputs a and b remain low, creating a decimal six.

In normal operation the lamp test and blanking are held high while store is connected to ground.

- Lamp test: If lamp test is grounded, all of the lettered outputs will go high.
- Blanking: If blanking input is made low, all lettered outputs go low, turning the display off.

Note as well that any input representing a number bigger than decimal nine will blank the display. Even though this 4-bit binary word can count from 0 (0000) to 15 (1111), the 4511 is designed to operate a seven-segment LED that cannot count above nine. Any binary word above 0101 is unreadable. The 4511 will blank, turning off all outputs to the display.

If the store input is made high, the value of the input code at that instant is held internally. With store high, the last value is held for display.

The 4511 BCD system layout is described in the Table L40-1.

Table L40-1   4511 BCD system layout

Input	Processor	Output
A binary word provides high signals at the binary inputs	Binary code is converted for display by the binary-counting decimal processor	Decimal value of binary input shown on the seven-segment LED

Breadboarding the Seven-Segment Display and 4511 Display Driver

Set the chip and seven-segment display near the end of your breadboard as shown in Figure L40-2.

Now build this circuit from the schematic in Figure L40-3.

The wire probes are only needed for testing in this lesson.

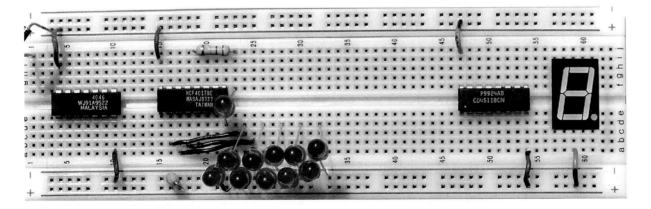

Figure L40-2

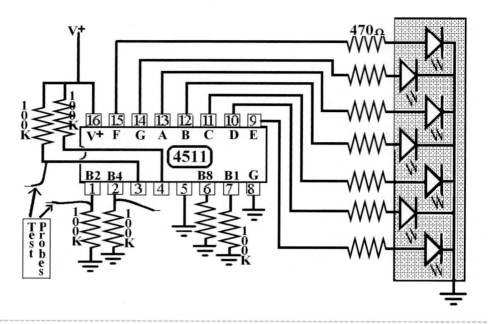

Figure L40-3

Building this circuit on the breadboard can prove to be very challenging. The picture in Figure L40-4 is provided to help guide you. There is a correct technique for breadboarding and then there is a mess.

- Note that the resistors have had their legs clipped shorter on one side. Then they are placed upright.

- Insulated wires are used to make the long connections. Those wires are a bit longer than the actual connection. That relieves any strain that might cause the wire to spring out of its hole.

- The wires are gently bent. Sharp corners actually encourage the solid wire to break under the insulation. Such breaks are frustrating and hard to find.

What to Expect

When you attach power to this circuit, the seven-segment display should read ZERO. That is, segments A, B, C, D, E, and F are lit. Not the crossbar, G. If you have a backward "6," you have switched the outputs "F" and "G."

1. Touch the wire probes connected from inputs B1, B2, B4, and B8 separately to V+. This "injects" a high signal into the various binary inputs.

2. Touch both B2 and B4 to voltage at the same time.

3. Try different combinations.

Figure L40-4

4. What happens when you try to inject a high at B8 and B2 at the same time? It should blank the display.

Troubleshooting

Run through the PCCP. The most common problems in this setup are as follows:

- Crossed wires.

- Reversed display.

- Individually burned-out LEDs on the display.

- Blanking at pin 4 and lamp test at pin 5 are connected to V+. Either connected to ground would keep this circuit from working.

Exercises: Control the Seven-Segment Display Using the 4511 BCD

The inputs B8, B4, B2, and B1 represent binary inputs.

Together, all four binary inputs create a 4-bit binary **word**. Figure L40-5 presents an excellent representation of how the binary word input relates to the display driver outputs.

1. Use the wire probe to connect B1 (pin 7) to voltage. What is the readout of the display?

   ____

2. Use the wire probe to connect B2 (pin 1) to voltage. What is the readout of the display?

   ____

3. Use the two separate wire probes to connect both B1 and B2 to voltage. What is displayed?

   ____

4. Work through Table L40-2.

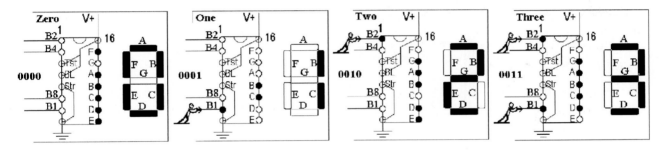

Figure L40-5

Table L40-2   Wire probe

Number	B8	B4	B2	B1	Binary Word	Seven-Segment Display Output
Zero	Low	Low	Low	Low	0000	A B C D E F
One	Low	Low	Low	High	0001	C and D
Two						
Three						
Four						
Five						
Six						
Seven						
Eight						
Nine						
Ten						

Lesson 40 — Control the Seven-Segment Display

The highest binary input that the 4511 recognizes is when inputs B8 and B1 are both high.

That means the highest it can count is 9. Anything over a binary input equal to "decimal 9" will cause it to blank out.

5. Why are inputs 1, 2, 6, and 7 connected to ground through 100k resistors? _____
_____

6. What is the maximum count of a 4-bit binary word? _____

7. What is the highest number that the 4511 seven-segment driver is able to display? _____

8. BCD stands for binary-counting decimal: Binary in–decimal out.

What is the binary word for decimal 7? _____

9. Besides the binary word input, there are three other inputs to the 4511. What are they (from the 4511 data sheet)?

a. _____ at pin_____ is conditioned _____ in normal operation. What is its purpose?

b. _____ at pin_____ is conditioned _____ in normal operation. What is its purpose?
_____

c. _____ at pin_____ is conditioned _____ in normal operation. What is its purpose?
_____

10. Test blanking and lamp test separately. Simply connect them to ground momentarily.

Lesson 41: Decimal to Binary—The 4516

The 4516 is used here. It changes the decimal clock signal from the 4046 VCO to a binary output used by the 4511. This lesson explains the operation and application of the 4516 BCD IC.

Data Sheet: 4516 Decimal-Counting Binary—Up/Down Zero to 15 Counter

The system diagram and pin assignment for the 4516 are shown in Figure L41-1.

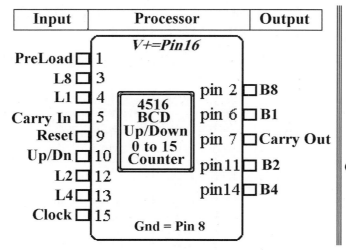

Figure L41-1

Basic Operation

For each clock signal input, the binary word output changes by one. The binary word shows as a high signal on corresponding outputs. For example, for the decimal number "7," the output B8 is low, while B4, B2, and B1 are high [0111].

In normal operation the inputs carry-in, reset, and preload are held low.

The binary word increments one count on the clock signal at pin 15. The movement is dependent on the setting of the up/down control. The output appears as the binary word at the outputs B8, B4, B2, B1.

Up/Down Control

When the U/D control is set to high, the count proceeds upward. When decimal fifteen [1111] is reached, the count cycles back to zero [0000] and carryout is triggered. When the U/D control is set to low, the count proceeds backward. When decimal zero [0000] is reached, the count cycles backward to fifteen [1111], and carryout is triggered.

Carry-In

The carry-in must be held low to allow counting. Holding the carry-in high stops the count, making it function like the enable switch of the 4511.

When multiple stages are used, the chips share the clock signal as shown in Figure L41-2. carryout of each stage connects to the carry-in of the next level.

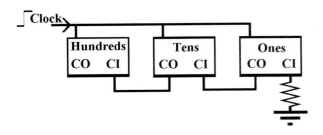

Figure L41-2

Reset

By momentarily connecting reset to V+, the count returns to zero [0000]. The reset must be returned to ground to allow counting to continue.

Preload

Any number can be preloaded into the 4516 by presetting the appropriate binary word to voltage through the preload inputs [L8, L4, L2, L1]. These preloaded inputs are loaded on to their corresponding binary outputs [B8, B4, B2, B1] by bringing the preload at pin 1 high for a moment. As with reset, the preload must be returned to ground to allow counting to continue. NOTE: Pin 15 should be in a low state and there should be no clock inputs to pin 15 when preloading is done.

Add to the System on the Breadboard

Leave a small patch of five free lines on one side for the last part of the system introduced in the next lesson.

As the schematic shows in Figure L41-3, the 4516 uses seven more 100-kilo-ohm resistors and more wire. You also add a temporary display of the binary word using four extra LEDs.

1. Remove the 100-kilo-ohm resistors from the 4511's inputs.

2. Preload control at pin 1 and the preload presets L8, L4, L2, L1 are attached to ground through 100-kilo-ohm resistors.

3. Carry-in is connected to ground, because it is an unused input.

4. Carryout can be left open. It is an output used to trigger another stage in counting.

5. Reset is attached to ground through a 100-K ohm resistor.

6. Start with the up/down control connected by a 100-kilo-ohm to V+.

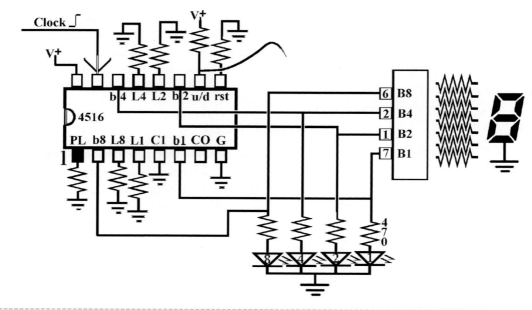

Figure L41-3

What to Expect

What to expect before you start to play. Use the clock signal from the carryout of the 4017 as the INPUT clock signal to the 4516. The number will advance one for every run down all ten LEDs. Note that this wiring is specific to where you are now. It is subject to being played with by curious fingers and nimble minds.

Also, at this point, the up/down control at pin 9 can be connected to V+ (counting up) or ground (counting down) through a 100-kilo-ohm resistor.

Exercises: Decimal To Binary—The 4516

A: Initial Setup

The 4516 acts as the interpreter. It changes the decimal clock signal source into the binary input used by the 4511 display driver. Use the partial schematic shown in Figure L41-3 to include the 4516 IC into your system diagram.

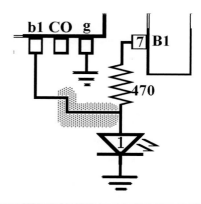

Figure L41-4 Common Error—Do not wire like this!

The outputs of the 4516, B8, B4, B2, and B1 are *wired directly* to corresponding inputs of the 4511. The four LEDs display the 4516's binary word. It matches the decimal number displayed on the seven-segment LED. Now place a 10-megohm resistor from pin 12 of the 4046 to ground. This sets the minimum frequency output of the 4046 to about 1 Hz. This way, you don't have to continually trigger the roll down.

Make sure that the 4516's up/down counter (pin 10) is conditioned to V+ so it counts upward.

2. The shortened "DCB" stands for what?

   D_____ C_____

   B_____

3. The shortened "BCD" stands for what?

   B_____ C_____

   D_____

B: Inputs Inputs Everywhere

Be careful while you are exploring. You can accidentally connect V+ directly to ground. That shorts out the power supply. Even though everything looks like it turns off, a short circuit can damage your power supply.

- Up/down counter. First, try out the up/down counter at pin 10. Use a long wire as shown in Figure L41-5 to connect the up/down control to ground.

- The reset. Use a wire probe and momentarily touch the reset input to voltage as shown in Figure L41-6.

- The preload. This is easy to do. Don't remove the resistors. They allow flexibility. Pop a

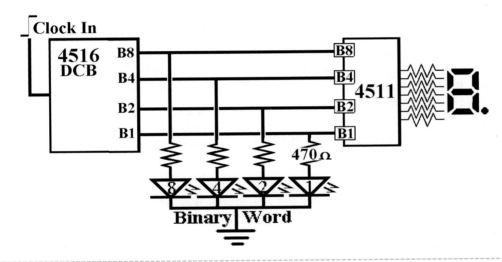

Figure L41-5

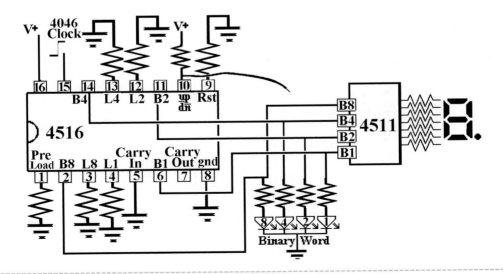

Figure L41-6

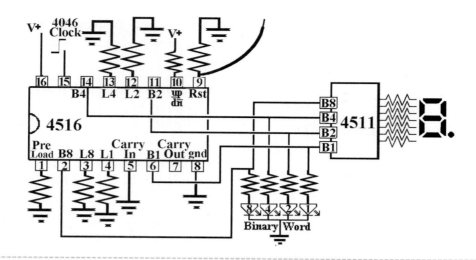

Figure L41-7

wire into both L2 and L4 as shown in Figure L41-7, and connect them to V+. Connect a long wire to preload. Touch the preload wire to V+, just for a moment. Your display should now show the number "6." When the preload is set to high, the values preset on the load inputs are dumped onto their matching binary count outputs.

C: Now for Some Serious Playtime! Exploring the Possibilities

Right now, both the 4017 and 4516 are connected directly to the clock output of the 4046 as shown in Figure L41-8.

1. The 4046 clocks the 4017. Now connect the 4017's carryout to the clock input of the 4516 as shown in Figure L41-9. Think about it. Try to explain what is happening.

2. This is even better. The 4046 clocks to the 4516. Connect the carryout of the 4516 directly to the clock input of the 4017. The setup is shown in Figure L41-10. What is happening?

3. Will the 4516 still count zero to nine in the setup above when you play with the reset or enable of the 4017? Try it out.

4. Remember the carryouts of the 4017? It is high for the first five number, then goes low for the second five numbers. Try the setup

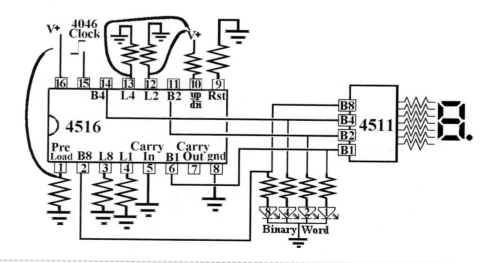

Figure L41-8

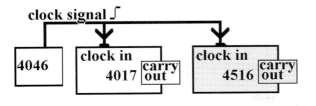

Figure L41-9

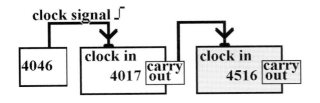

Figure L41-10

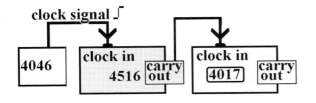

Figure L41-11

shown in Figure L41-11. Connect the 4017 carryout to the up/down counter of the 4516. Predict what's going to happen. Were you right?

5. Any other ideas?

D. Setting Up the DigiDice

Figure L41-12 is the graphic system diagram that you can use to set up your prototype for DigiDice. Notice that some particular features have been incorporated into the system.

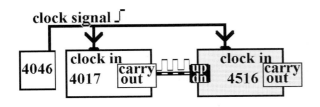

Figure L41-12

The 4046 clock feeds to both 4017 and 4516.

When "Out 6" of the 4017 goes high, this signal is used to trigger the following:

1. The reset of the 4017 (the first six LEDs light up in sequence)

2. The preload (pin 1) of the 4516

   - L1 (pin 4) must be set directly to voltage.
   - This will load the number "1" when it is triggered, effectively skipping "0."
   - Resulting in numbers displayed 1-2-3-4-5-6-1-2-3 . . .

At this time, you can remove both the 10-megohm resistor from the 4046's pin 12 and the LEDs used to display the binary word. The minimum frequency will return to full stop, and you will have room for one last addition to the entire system.

Lesson 42: Automatic Display Fade-Out

Hey! If you didn't do it yet, do it now. Remove those LEDs used to show the binary word.

Power consumption is important if you use batteries to power the system. CMOS Electronics are popular because the systems use so little power. In comparison, the LEDs gobble energy in this system. In this lesson we apply a simple RC to the system that automatically cuts the power to the LEDs.

A basic system diagram is shown in Table L42-1. It includes the RC that automatically turns off the LEDs.

Table L42-1 Basic system diagram

Input	Processor 1	Processor 2	Processor 3A	Processor 4	Output 1
A contact switch like a PBNO	RC1 controls the timing of the rolldown	4046 VCO RC2 controls max and min frequency of the clock signal	4017 Walking ring 1-10 counter. Seventh LED (out 6) is connected to reset and preload controls on the 4516	Timed off as a power saver. RC3 controls the transistor that cuts power to the LEDs	6 LEDs

			Processor 3B		Output 2
			4516DCB ⑨ 4511BCD		Seven-segment display counting 1-2-3-4-5-6

A detailed system diagram for DigiDice deals with the input/processor/output of each piece of the system, Table L42-2.

Table L42-2 Detailed system diagram

First Section, Including the 4046

Input	Processor	Output
Push button closes Voltage filling RC1	Falling voltage controlled by the RC1 drains from max to minimum. The voltage at pin 9 is compared to V+. This controls speed of voltage-controlled oscillator	Length of rolldown controlled by RC1. Rolldown clock signal.

Second Section 4017

Input	Processor	Output
• Clock rolldown from the output of the 4046 • Reset	1 of 10 counter reset at out 6.	• Walking ring provides high to out 0 to out 6. • Out 6 is used to reset 4017 to 0. • Out 6 is used as controlling input to 4510.

Second Section LEDs

Input	Processor	Output
Walking ring provides high to 1 of 6 LEDs in sequence.		Light

Third Section Starting with 4516

Input	Processor	Output
• Clock rolldown from 4046 • High input from 4017 out 6 triggers preload of L1 • U/D counter set up or down	Decimal counting binary	Advancing binary code

Third Section 4511

Input	Processor	Output
Binary word	Converter for seven-segment LED	Decimal value of binary input shown on the seven-segment LED
Binary counting Decimal		

Fourth Section Timed Off

Input	Processor	Output
Voltage filling RC3	Transistor opens, then slowly closes the connection to ground as controlled by RC3	Displays fade-out

Displaying the system this way can be confusing and long-winded. A "graphics system diagram" shows how all of that information can be displayed more efficiently. Such diagrams are more readily understood and infinitely easier to use when designing and troubleshooting systems. Figure L42-1 is the graphic system diagram showing how RC3 is set into the system.

But even a graphic system diagram does not show the full schematic. For that matter, at this point, I'm not going to show the full schematic either. All that needs to be shown is the specific schematic of interest and the notation of how it is connected into the system. Figure L42-2 is the schematic showing how RC3 controls the NPN transistor.

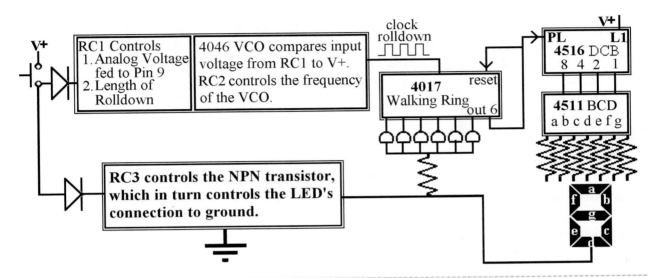

Figure L42-1

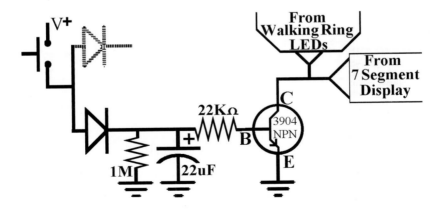

Figure L42-2

Everything used here should be fairly obvious, except for the extra diode. Why is the extra diode there? The diodes separate RC1 from RC3. Here is a complete explanation why it is necessary.

RC1 and RC3 are isolated from each other using the diodes to prevent any reverse flow of the current. If there were no diodes, the charges would be shared between the capacitors of RC1 and RC3. This is shown in Figure L42-3.

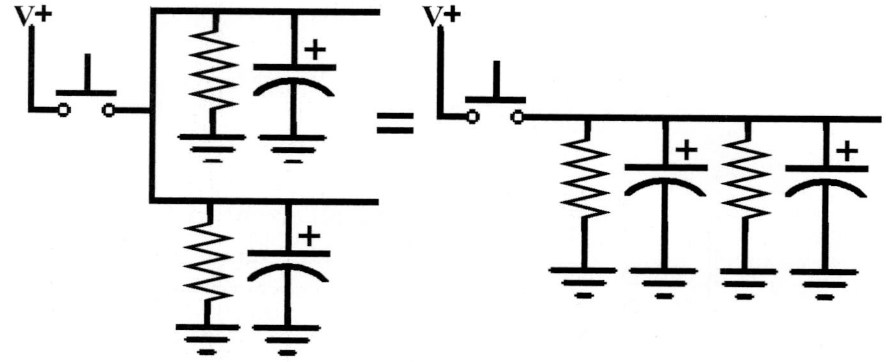

Figure L42-3

The resistors would act together, too. RC1 and RC3 would combine and act as a single RC.

If there were only one diode, the unisolated capacitor shares its charge first. This is demonstrated in Figure L42-4. The charge in the 1-μF capacitor of RC1 would push through the lower diode and add to the charge of the 22-microfarad capacitor of RC3. The reverse would also occur; if the bottom diode were not present, the charge stored in the 22-μF capacitor of RC3 would push through the upper diode and add to the charge of the 1-μF capacitor of RC1.

The complete and proper setup is shown in Figure L42-5. Each RC acts independently. The RC1 rolldown should finish 5 to 10 seconds before the fade-out really finishes. If the LEDs fade out before the rolldown is completed, adjust the timing of either RC1 to be shorter or the timing of RC3 to be longer.

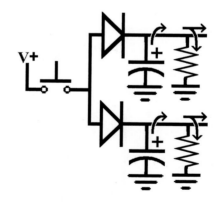

Figure L42-5

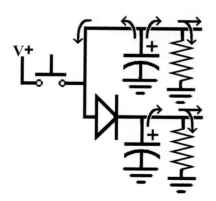

Figure L42-4

Define, Design, and Make Your Own Project

It has been my experience that when people are given instructions to make something for themselves, they will complain. Either they are given far too much detail or what they perceive as not nearly enough. All this from the same audience. What I have done here is give examples. The more complex the example, the more I choose to talk in terms of concept, not detail. I do this on purpose. This is your project. If I gave you all of the details, it becomes my project, and you copying my project.

Remember the earlier comment about Lego. That was a serious statement. Parts of one fit perfectly into the parts of another. That is one of the beauties of digital electronics. Think about it. You have learned enough that you could almost build the controller for an arena's score board. You can develop your own applications using just a portion of the system. Next time you go to the mall, carefully look at those fancy gadgets that eat your quarters and tell you your fortune. What? You say you can't do that? Then why not?

Lesson 43: Defining and Designing Your Project

Ideas to keep in mind:

1. In an ideal world, the sky is the limit.

2. Murphy was probably right. Murphy's first law states, "If anything can go wrong, it will." Do an Internet search of "Murphy's law" + Technology.

3. In reality you will never have
   - Enough time
   - Enough money
   - The right equipment for the job

4. Your expectations for this project should be reasonable.
   - Five printed circuit boards are provided in the kit for this unit. This provides you with maximum flexibility.
   - The biggest limitation that exists for this project is your inability to create your own enclosure.
   - The most reasonable and inexpensive pre-made enclosure is the plastic VHS cassette case.
   - These are often available with plastic lining that allows you to insert labels, signs, and directions.

Notes Regarding Possibilities

This is digital electronics. The processors can be mixed and matched, or not used at all.

1. 4046 VCO
   - Can be set to a specific frequency
   - Rolls down
   - Can roll up (think about it)
   - Can set minimum frequency
   - Can set maximum frequency
   - Has frequency range

2. 4017 walking ring

- Dependent on a clock signal, but any clock signal will do.

- Don't underestimate reset or enable.

- Can count zero to nine, and reset or freeze on any predetermined number.

- What about the carryout? High for half the count, low the other half.

- What about using two 4017s, counting 0 to 99 in rows.

- The outputs are not limited to LEDs as shown in Figure L43-1. They can control transistors to give power.

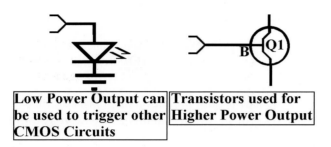

| Low Power Output can be used to trigger other CMOS Circuits | Transistors used for Higher Power Output |

Figure L43-1

3. 4516 decimal counting binary

- Like the 4017, it is dependent on a clock signal.

- With a 1-Hz signal, you could make a binary clock.

- The 4516 has more inputs than it has outputs.

- The U/D control takes any digital input. It is not dependent on a clock signal.

- What about having two 4516's counting 0 to 99 in digits?

Timing

There are only three separate RC timers right now. More are possible. RC1 and RC3 work together. If RC1 is longer than RC3, the displays fade before the rolldown is complete. Use Table L43-1 to set the timing of the RCs to your desires.

Table L43-1   Set timing of the RCs

Resistor	Capacitor	Time
20 MΩ	10 μF	200 seconds
10 MΩ	10 μF	100 seconds
4.7 MΩ	10 μF	50 seconds
20 MΩ	1 μF	20 seconds
10 MΩ	1 μF	10 seconds
4.7 MΩ	1 μF	5 seconds

Examples

Each of five examples presented has a unique twist in its application. New portions are explained in detail. Some of the design components, by their very nature, are identical between projects. When that happens, that portion is mentioned briefly.

The Ray Gun

The ray gun displayed in Figure L43-2 is a wonderful little project.

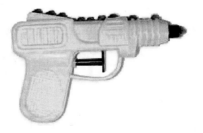

Figure L43-2

The entire system has been enclosed in a squirt gun. The unique idea here is that only a portion of the larger system has been applied as shown in the graphics system diagram of Figure L43-3.

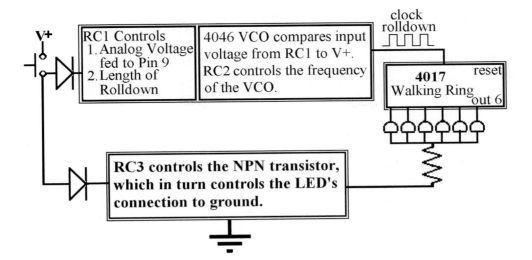

V+

RC1 Controls
1. Analog Voltage fed to Pin 9
2. Length of Rolldown

4046 VCO compares input voltage from RC1 to V+. RC2 controls the frequency of the VCO.

clock rolldown

4017 Walking Ring

reset

out 6

RC3 controls the NPN transistor, which in turn controls the LED's connection to ground.

Figure L43-3

The same system can be applied to a push button fortune teller or a 10-LED roulette wheel.

The "Whatever ?" Detector

The only unique quality here is the switch and the name you give it. In fact, this is the same system, but applied as a fake metering device. Fill in the word for "whatever." The beauty is that it actually changes frequency from person to person. Did you notice that you could get the circuit to work by just touching across the bottom of the push button? Try it. The packaging can be very fancy, as shown in Figure L43-4, or you could use a simple box.

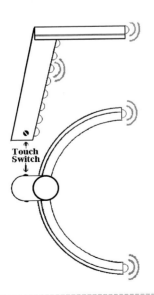

Touch Switch

Figure L43-4

You could moisten your palm and pick up the "Ghost Detector," claiming all sorts of things as you point it toward a dark corner. Give it to your friend and watch as it goes dead. A little knowledge isn't dangerous. It's fun.

An Animated Sign

The animated sign shown in Figure L43-5 could be used for a variety of shapes.

Dew Drop Inn

Figure L43-5

Both of the things that make this unique are shown in the graphics system diagram of Figure L43-6.

First is the subsystem that triggers the 4046. It is a 4011 oscillator set to pulse every 20 seconds. The maximum frequency is set to 20 hertz. It rolls down almost to a complete stop before it kicks the oscillator high again. The schematic is shown in Figure L43-7.

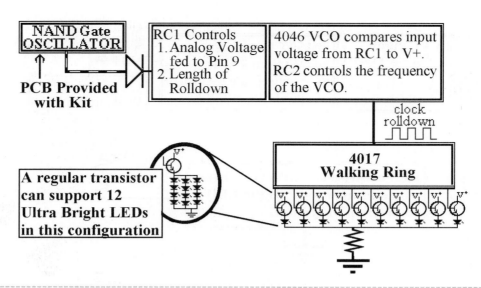

Figure L43-6

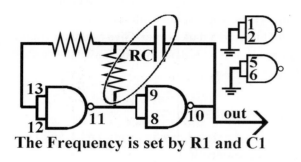

The Frequency is set by R1 and C1

Figure L43-7

Second is the necessity for ultrabright LEDs to make the output visible from a distance. This requires the output of the 4017 to use transistors to amplify the power.

The IQ Meter

This is another wonderful gadget. It could just as easily be called a lottery number generator. You secretly control the first digit by a hidden touch switch made from two pin heads. Shown in Figure L43-8, when you hold it, you're guaranteed a number above 100.

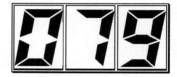

Figure L43-8

Everybody else gets a number under 100. It doesn't use the walking ring at all. To do it properly, though, it does need two sets of the number display circuits and three seven-segment LEDs. Directions to make your own printed circuit boards are provided at wwwbooks.mcgraw-hill.com/authors/cutcher. Figure L43-9 shows the graphic system diagram for the IQ meter.

Modify the dual NAND gate PCB, included in the kit, so it is a single touch switch. You can figure out the wiring to get a high output to show one and a low output to show zero on the seven-segment display.

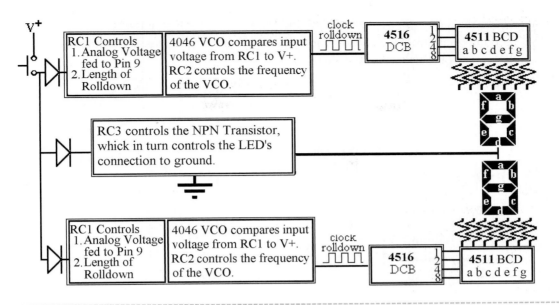

Figure L43-9

Love Meter Fortune Teller

This wonderful gadget is popular in the malls. It takes two inputs to work. You pay a quarter and do one of several different things. It might be that you put your palm on the outline of a hand or pass your hand over the hand of the plastic gypsy. You might touch one pad and your girlfriend touches the other; then you kiss. Then the lights flash, and you get your fortune told. The outcomes shown in Figure L43-10 come directly from The Magic 8-Ball® toy made by Tyco Toys.

The graphic system diagram for the fortune teller is shown in Figure L43-11.

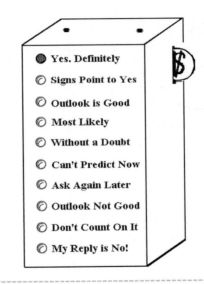

Figure L43-10

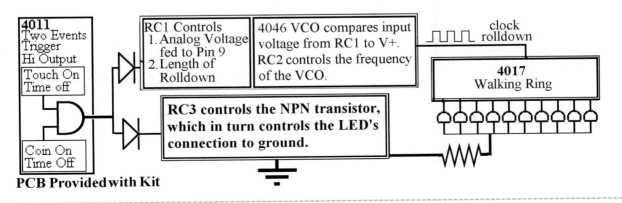

Figure L43-11

The schematic for the dual input trigger is shown in Figure L43-12. It is important to give plenty of time for the coin input to time out, so that by the time they trigger the second input, the first has not timed out.

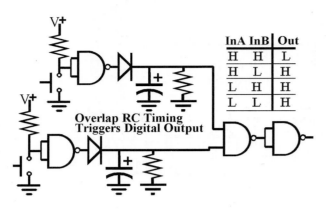

InA	InB	Out
H	H	L
H	L	H
L	H	H
L	L	H

Overlap RC Timing Triggers Digital Output

Figure L43-12

Event Counter and Trigger

This device will count how many times something happens, and then trigger an output. With two 4516 chips, the output can be set to trigger at any number between 0 and 255. The graphic system diagram in Figure L43-13 is set up for a single numbering system.

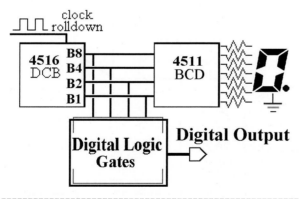

Figure L43-13

The 4516 can be preset to any number zero to 15. You don't have to feel confined to what the display can show. Set the up/down control to count backward. The output is triggered when everything

reaches zero. Figure L43-14 shows how to apply two 4071 quad OR gates to check to logic output.

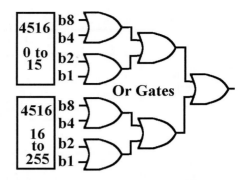

Or Gates

Figure L43-14 *The output is low only when input is 00000000.*

The 0000 0000 can be made by the following:

- Counting up and rolling over the top number
- Counting down until you hit the bottom
- Triggering reset by an outside event

The Slot Machine

It is possible to do this. What if I get two rows of walking rings and trigger them both? When the LEDs match, it can work like a slot machine and trigger a payout. What a simple concept. Move forward with this one only if you have confidence. It can work, but it takes work. The graphic system diagram for a two-line slot machine is shown in Figure L43-15.

It appears obvious that you can compare the outputs using an AND gate. The 4081 dual input quad AND gate is ideal. But consider that a high output from the two walking rings might cross each other for a millisecond. That is enough to trigger the AND gate falsely. It is important to slow down the comparison to prevent false triggering. Such a circuit is displayed in Figure L43-16. This ensures that each LED in the pair must be high for a minimum of 1/2 second before the AND gate input moves above the triggering voltage.

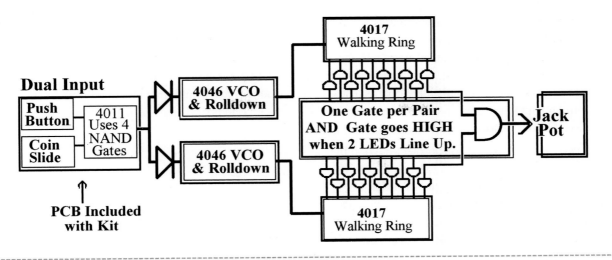

Figure L43-15

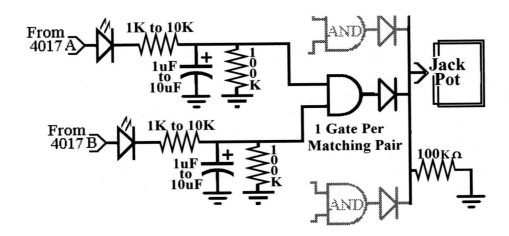

Figure L43-16

What about the payout? I'll be honest. Check your phone directory for any shop that repairs vending machines. Unless you are a skilled machinist and have time, you will save yourself many headaches by purchasing a used coin return mechanism. Usually, one clock signal in kicks one quarter out. It would be fairly safe to say the output from the 4081 will be a clean clock signal. Have fun.

One Last Word

Be creative. Be patient with yourself. Breadboard your prototype before you start building onto your printed circuit board.

Lesson 44: Your Project: If You Can Define It, You Can Make It!

Lesson 44 includes the following for building each subsystem of your project.

- The printed circuit board layout
- The schematic laid out onto the PCB
- The parts placement onto the PCB

The possibilities for creating different applications with the subunits of this circuit should be limited only by your imagination. But I have to apologize. There are real limitations to this course. I have strived to provide you with a wide range of printed circuit boards. You can control all of the various inputs for each processor. Be reasonable and realistic in your applications.

If you choose to expand your project beyond the components provided, they should be available at any "electronics components" store or order them from www.abra-electronics.com. If you really want to play, there are instructions for creating your own printed circuit boards at www.books.mcgraw-hill.com /authors/cutcher.

WIRE TYPE: A comment is necessary regarding the type of wire that you are using as you start to build your project. Up to this point, you have been using 22- to 24-gauge solid core wire for prototyping on your breadboard. Solid wire is great for prototyping. But it breaks easily. Bend a piece back and forth 10 times. If it didn't already break, it will soon. As you build your subsystems, use stranded wire to connect them. Stranded wire will bend back and forth without straining itself. But it is difficult to use on breadboards. This wire is readily available at dollar stores, but disguised as "telephone wire."

Inputs

There are four different types of input switches.

No Modification

- The contact switch: Recall that basic mechanical switches were discussed in some detail in Part II.

- The touch switch: The initial switch using the zener diode is sensitive enough to react to skin's resistance. This is not digital. Connect the push button's wires to pin heads instead.

Two Input Digital

A coin or other action at the SW1 in Figure L44-1 will turn the output of the first gate high for about 20 seconds. You would then have that time to activate the second NAND gate using the SW2.

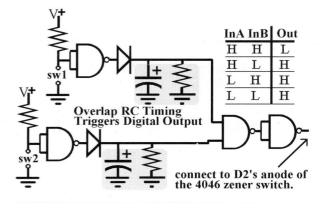

InA	InB	Out
H	H	L
H	L	H
L	H	H
L	L	H

Overlap RC Timing Triggers Digital Output

connect to D2's anode of the 4046 zener switch.

Figure L44-1

The bottom view of the PCB is shown next to the top view parts layout drawing in Figure L44-2.

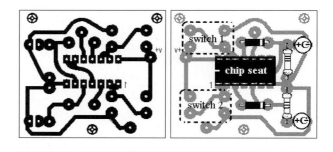

Figure L44-2   *Note that the output is at pin 10.*

Many of the possible variations for these input switches are discussed in depth in Part II. Use the same layouts and component values depending on the type of switch you want to build.

The Self-Kicking Oscillator

The schematic for the self-kicking oscillator is shown in Figure L44-3. Remember that the frequency is set by the RC. The inputs of the unused gates have been set to ground.

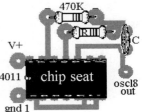
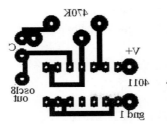

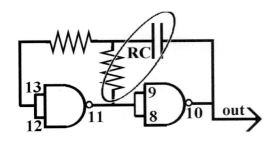

Figure L44-3

The timing for the rolldown can be adjusted, but the oscillation timing should be longer than the rolldown timing if you want the rolldown to stop completely before it gets kicked again. The schematic is shown in Figure L44-4.

Processors

VCO and Timed Off

The heart of the system's rolldown is the VCO's reaction to the voltage drop of RC1. To save space, three subsystems have been grouped together on one printed circuit board. Naturally, RC1 and the 4046 VCO would inhabit the same board. I've included the Timed Off circuit on this board as well. Figure L44-5 shows the individual schematics that exist on this one PCB.

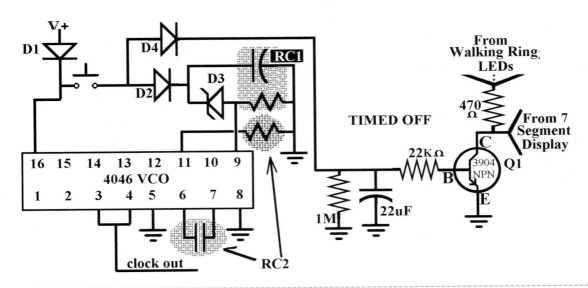

Figure L44-5

This does make for a more complex PCB layout, as you can see in Figure L44-6. Note that the components that make up the Timed Off portion have been marked separately.

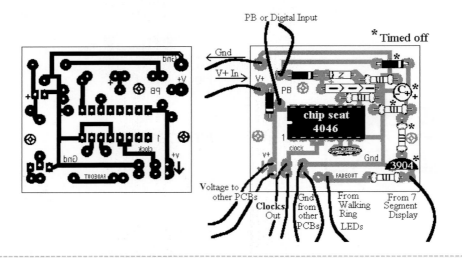

Because the PCB is fairly crowded, another view in Figure L44-7 is provided. This top view identifies each component by value and relation.

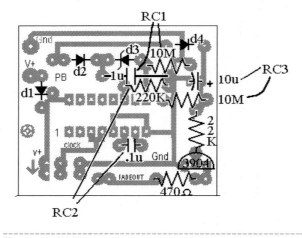

Figure L44-7

The Walking Ring

The schematic of the 4017 walking ring in Figure L44-8 says it all.

Remember that any digital high can be used to control the reset or enable. Otherwise, these inputs

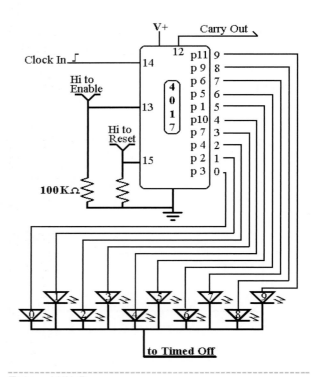

Figure L44-8

need to be held low. Figure L44-9 shows the PCB and parts layout.

Even thought this is not as complex as the previous PCB, Figure L44-10 emphasizes the extra control inputs available on the walking ring PCB.

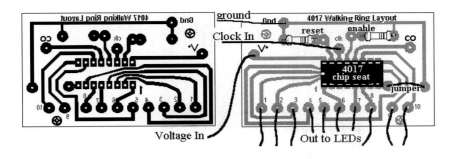

Figure L44-9

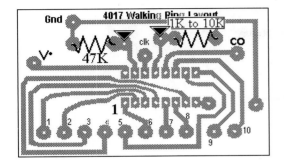

Figure L44-10

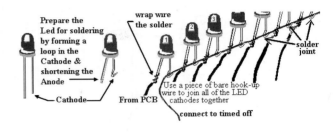

Figure L44-11

The LEDs are not expected to live on this PCB. Remember to use stranded wire to connect the LEDs. Also, there is no ground connection for the LEDs on the 4017 PCB. That is because each of the LEDs negative leg (cathode) is wired together as shown in Figure L44-11 and that one connection leads back to the Timed Off circuit.

So now you have 10 or so LEDs floating on wires. They can be neatly mounted onto an enclosure using the LED collars. The collars fit snugly into holes made by a 7-mm or 9/32" drill bit. Figure L44-12 shows how to use these marvelous little items.

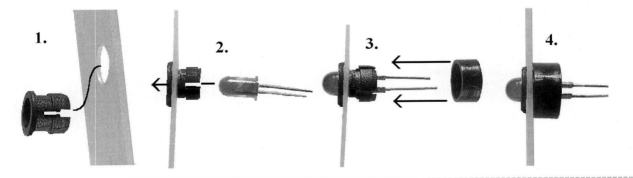

Figure L44-12

4516/4511 and Number Display

Examine the schematic in Figure L44-13.

Inputs Inputs Everywhere

Consider all of the preloads as a single input. They have to act together, and are controlled by pin 1 of

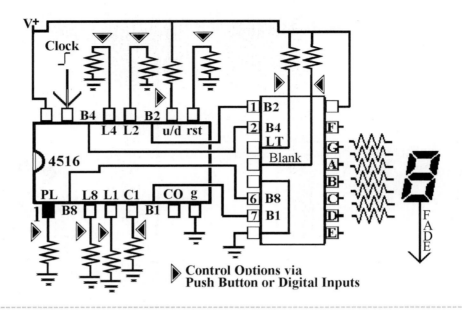

Figure L44-13

the 4516. Also, ignore carry-in unless another 4516 feeds to it. So there are really only five inputs on this subsystem.

The printed circuit board layout and parts placement are all displayed nicely in Figure L44-14.

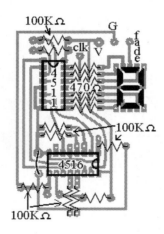

Figure L44-15

Remember to use stranded wire for all of the connections to the other subsystems. All of the resistors are marked either 100 kilo-ohms or 470 ohms in Figure L44-15. These are necessary to condition all of the many inputs that exist.

That's really all there is to it. The rest is up to you.

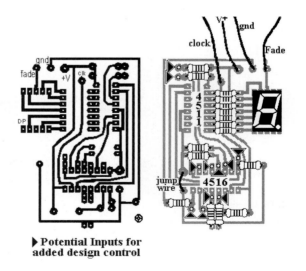

▶ Potential Inputs for added design control

Figure L44-14

Lesson 44 – Your Project

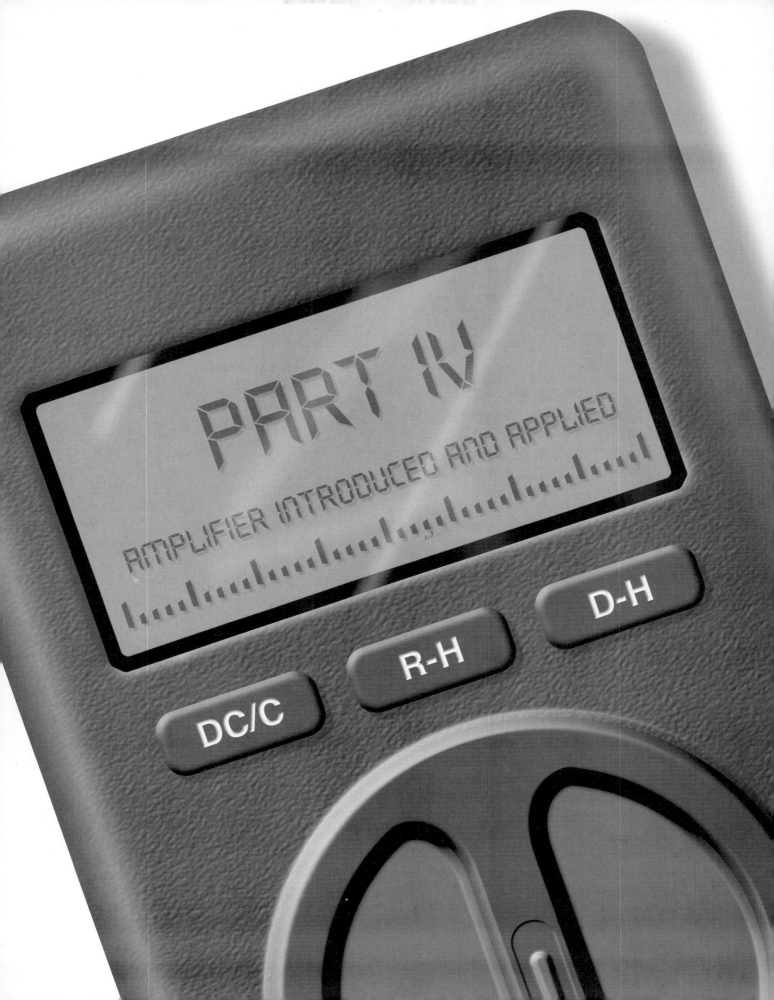

PART IV

AMPLIFIER INTRODUCED AND APPLIED

What Is an Amplifier?

Lesson 45: Transistors as Amplifiers and Defining Current

So, what is an amplifier? It takes a small signal and makes it bigger. Signals are made of both voltage and current. We've applied the transistors as amplifiers in a few ways. You may not have recognized it at the time. Let's review what you did.

The definition from dictionary.com is very simple.

> **am·pli·fy**—verb. 1. To make larger or more powerful; increase. 2. To add to, as by illustrations; make complete. 3. To exaggerate. 4. Electronics—To produce amplification of or to amplify an electrical signal.
>
> **am·pli·fi·er**—noun. 1. One that amplifies, enlarges, or extends. 2. Electronics—A device, especially one using transistors or electron tubes, that produces amplification of an electrical signal.

Review

A. Remember the Night Light

Its action is shown in Figure L45-1. The signal given to the base of the NPN transistor is amplified.

The voltage divider created by the potentiometer and the LDR control the voltage to the base of the transistor. Remember, the voltage at that point is dependent on the following:

1. The setting of the potentiometer

2. The amount of light on the LDR

The signal to the base of the transistor has little power, not even enough to turn on an LED, because so little current gets through the 100-kilo-ohm potentiometer set at midrange.

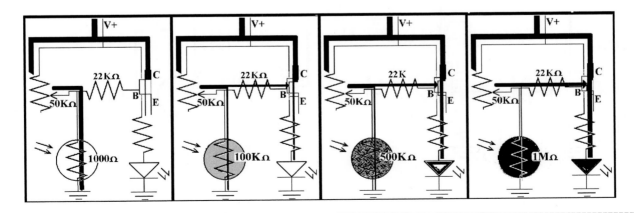

Figure L45-1

The action of how the signal to the base is amplified by the NPN transistor is demonstrated in Figure L45-2. This action acts as a valve controlling the power source to the LEDs:

- The smaller signal controls the transistor's valve action.

- The resulting amplified signal from C to E is a much more powerful version of the original small signal.

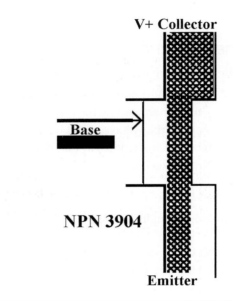

Figure L45-2

B. Audio for the NAND Gate Oscillator

Remember when you first connected your speaker directly to the NAND gate output. The signal was so quiet that you had to place your ear right on top of the speaker to hear it . . . and that was in a quiet room. The amount of current passing from V+ to pin 10 when it was low was enough to light an LED, but not enough to really shake the speaker. This is the schematic in the left side of Figure L45-3.

That small signal to the base was amplified by the PNP transistor, shown inserted into the schematic in the right side of Figure L45-3. The small signal was used to control the transistor valve action that controlled the power directly from the voltage source.

The weak signal to the base was amplified by the PNP transistor, providing plenty of power to the speaker coil to make an amazingly annoying loud sound.

"Wait a minute . . . ," you say. "Didn't the digital gate provide 9 volts, and the transistor gave 9 volts as well, so why was the transistor output more powerful?"

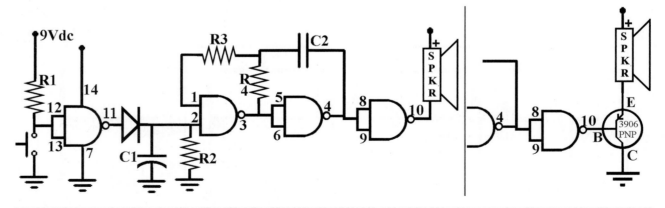

Figure L45-3

Calculating the Current: Amperage

The difference was not in the voltage, the push behind the electrons, but in the quantity of electrons being pushed, the current.

Figure L45-4 starts a very practical way of picturing *current in an electrical system*.

Figure L45-5

Figure L45-4

In electronics, voltage is the push behind the flow, or the force.

For a creek, gravity is the push, the force behind the flow, as water runs downhill.

A creek, by definition, has little current. Current is measured by how quickly water passes by in Liters per second or cubic feet per second. This creek has a current of 5 ft³/s.

> cur·rent (noun). 1. A steady, smooth onward movement: a current of air from a fan; a current of spoken words. Synonyms—flow. 2. The part of a body of liquid or gas that has a continuous onward movement: rowed out into the river's swift current. 3. A general tendency, movement, or course. Synonyms—tendency. 4. Symbol I, Impedance (a) A flow of electric charge. (b) The amount of electric charge flowing past a specified circuit point per unit time. (dictionary.com)

Now think of a stream like the one pictured here in Figure L45-5. It has a modest current, but you could walk across it.

Consider that the downhill "slope" is the same as the creek. The force is the same, but the big difference is the "current," the amount of water. It is more powerful.

This stream has a current of 100 ft³/s.

This river pictured in Figure L45-6 has a very large current. Even though it has less "slope" as the creek and stream shown above, it has more water flowing by a single point, every second. If you tried to swim across, you would be swept away.

Figure L45-6

Obviously, the river is the most powerful. The force is the same, but the difference here is the amount of water. The river has a current of 20,000 ft³/s.

Current in electricity is electrons passing through a wire.

The standard unit of current is an ampere (Amp or A, for short): 1 Amp = 1 Coulomb/second (1 Coulomb = 10^{18} electrons)

Current is the general term. Amperage is the unit of current. The common abbreviation for current is *I* (impedance). People new to electronics say "It's such a pile of new words. 'I' looks odd." But think of the word *stampede* with a herd of electrons pounding through the wires. Now that's an "impede." There's a wild one shown in Figure L45-7.

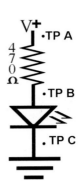

Figure L45-7

So how much is a coulomb? How much current is moving through an electronic system? How could we look at the electrical equivalent to a creek in Figure L45-8, and figure the current used by an LED?

The current is a quantity we cannot measure directly with the tools at hand. But we can figure the amperage by using "Ohm's law." Ohm's law simply

Figure L45-8   *The electrical equivalent to a creek*

states that V = IR, where George Ohm calculated that the amount of pressure in a circuit is directly related to the amount of current passing through the load.

Unit	Measuring
V = Voltage	Voltage is the unit of force.
I = Amperage	Amperage is the unit of current (impedance).
R = Ω	Ohms is the unit of resistance.

Here's how Ohm's law works. Think of a hose. A regular garden hose will do. The hose in Figure L45-9 has a fixed diameter, so the resistance is not going to change.

Figure L45-9

The higher the pressure, the more water flows through it. The lower the pressure, the smaller the current becomes.

But if you change the resistance as demonstrated in Figure L45-10 and Figure L45-11, that affects the amount of current, too.

Figure L45-10

Figure L45-11

The fire hose shown in Figure L45-12 offers very low resistance because the hose is bigger. It is able to handle lots more current, but the pressure from the city supply is the same.

Figure L45-12

Back to Ohm's Law (V = IR)

This is a simple three-variable equation: $A = B \times C$. If you know two variables, you can figure the third. But how would you figure the current in the simple LED circuit shown in Figure L45-8?

To figure the current flowing through the LED's circuit, you need to know (a) the voltage used by the resistor (the pressure drop) and (b) the resistance.

Now, back to considering the basic circuit. Let's make some assumptions.

1. The power supply is a perfect 9 volts.

2. "R": The resistor's value is a perfect 470 ohms.

3. "V": Connect power and measure the voltage drop across the resistor—from TP a to TP b. The voltage drop across the resistor is 7.23 volts. $V = IR$ so

$$\frac{V}{R} = I$$
$$7.23 \text{ V} = I \times 470 \text{ }\Omega$$
$$I = 0.0153 \text{ A}$$

So we just figure out how much current passed through the resistor. Does that tell us how much passes through the LED? Yes it does. Consider the following.

A simple statement. Unless some is added or taken away,

- The current in a hose is the same along the entire hose.

- The current in a creek is the same along the entire creek.

- The current in a simple circuit is the same throughout the entire circuit.

For a simple circuit, the amount of current is the same throughout the system. The current (I) passing through the resistor is the same amount of current through the LED as well.

Exercise: Transistors as Amplifiers and Defining Current

You can check your answers at www.books.mcgraw-hill.com/authors/cutcher.

1. Define amplifier as it relates to electronics.

   _____

2. Match these units to the terms they represent.

V, A, W, hose width, force, coulombs per second, gravity, ft³/sec, I
_____

Pressure	Current	Resistance

3. Define the two combined factors that make a current.

   a. _____

   b. _____

4. What is a basic unit used to show current in the following?

   Water System _____

   Electrical System _____

5. Write without scientific notation how many electronics there are in a 1-coulomb charge.

   _____

6. There are 0.0153 amperes in the LED circuit. Write out exactly how many electrons pass a single point in the wire in 1 second.

   _____ = 1.53 10¹⁶ electrons

Recognize that this is considered a very small current.

7. Now it is time to observe the actual effect that different resistors have on the current.

You might recall Figure L45-13 as the schematic Lesson 5, a long time ago. At that time, you observed what happened as you changed resistors. Now you get to understand what happened.

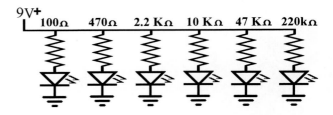

Figure L45-13

Do all figuring here to three significant figures. This does not mean figure to the hundredths place. It means that in terms of accuracy, all numbers have two digits. For example, a 10-kilo-ohm resistor has two significant figures because the color coding only shows two digits. On the DMM it might show 9.96 kilo-ohms. That shows three significant figures.

You can copy most of the information from the exercise you did for Lesson 5 to Table L45-1.

Table L45-1 Information from the exercise

Resistor in Order	DMM Resistor Value	Voltage Drop Across Resistor	Current in System Amps = $\frac{V_{drop}}{\Omega}$	Voltage Drop Across the LED
100 Ω	____Ω	____Volts	0.__ A	___Volts
470 Ω	____Ω	____Volts	0.__ A	___Volts
2.2 kΩ	____Ω	____Volts	0.__ A	___Volts
10 kΩ	____Ω	____Volts	0.__ A	___Volts
47 kΩ	____Ω	____Volts	0.__ A	___Volts
220 kΩ	____Ω	____Volts	0.__ A	___Volts

8. Think of your garden hose.

   a. If you turn the hose on, the pressure comes from where?_____

   b. Pressure in an electronic system is called what? _____

   c. If you leave the pressure the same, and squeeze the hose, this increases what?

      _____

   d. As you increase the resistance, what happens to the current? _____

   e. If voltage in a system remains constant, current will decrease if resistance is what?

      _____

   f. Conversely, the current will increase when resistance is what? _____

Figure L46-1

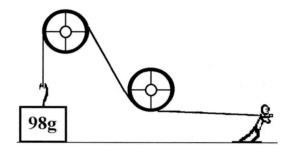

Figure L46-2

Lesson 46: Defining Work, Force, and Power

What are we amplifying? Here you learn how work, force, and power are defined and measured in electronics. Amplifiers by their very nature are analog devices. The transistors you've worked with are analog. They responded to any voltage to the base in varying degrees. Remember, NPN was turned on by more voltage to the base, and the PNP was turned off as the voltage increased. Here you will be introduced to the two types of amplifiers that match the opposite actions of the transistors you've already dealt with.

What Is Force

Force is the amount of energy exerted. In Figure L46-1 Atlas is exerting a force to hold up the sky.

But by definition, no work is happening because there is no movement. If an object is not moved, no work is done, no matter how great the force.

Force is measured in newtons. Roughly, 1 newton will move about 100 grams' mass upward against the

force of gravity as demonstrated in Figure L46-2. Specifically, 98 grams, but 100 is easier to remember.

The force in electricity is measured in volts. The matter being moved is electrons. This is shown in Figure L46-3.

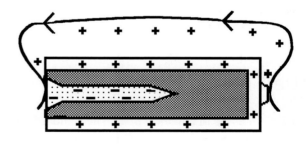

Figure L46-3

What Is Work?

Work is measured as the amount of force exerted on an object through a distance. There has to be both force and distance for work to be done. Work is measured in standard units of newton-meters or joules. 1 N-m = 1 J. The terms are interchangeable.

The distance can be done via a bottle rocket or a snail as shown in Figure L46-4. The speed doesn't matter.

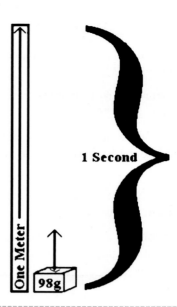

Figure L46-5

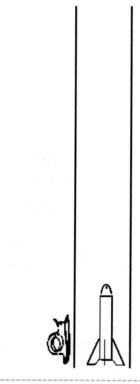

Figure L46-4

Watt Is Power?

Power is the *rate* that *work* is *done*. The standard unit of power in electricity is watts. One watt is the standard power unit defined by the force needed to move 98 g upward against gravity 1 meter in 1 second. This is graphically shown in Figure L46-5.

Two items can have the same force exerted on them, but can have different power. The bottle rocket may do 1 joule of work in 0.01 second. That works

out to 100 watts. The snail may take 1,000 seconds to do the same work. It has a power of 0.001 W.

But look at the power in different water systems. The force of gravity pulls water over the edge of the cliff in both of the waterfalls shown in Figure L46-6. The current of both waterfalls has the same force of gravity on them. It is 180 feet at each of these falls. But power is defined by the current multiplied by the force. By definition, the current includes both quantity and rate. I would stand under only one of them.

Figure L46-6

Multiply the current by the voltage and that gives power. In any system, power is determined by how fast work is done.

In these natural water systems

Power = current × force.

Current = ft3/s = the amount of water passing by and how fast it is flowing.

Force = slope = the amount of push behind the current, the pressure.

In electrical systems

Power = current × force.

Force = voltage.

Current = amperage measured as a standard where 1 A = 1 C/s.

Remember that 1 C is a standard mass of electrons. Power is measured in watts.

1 W = 1 A × 1 V

The LED is a low power light output. How much power is used by the LED? Consider the basic circuit shown here in Figure L46-7.

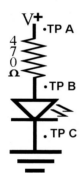

Figure L46-7

To figure the power used by the LED, you need to know both voltage and current.

Now this becomes a three-step process. But they are three simple steps.

1. You need to measure the voltage drops across the LED (TP b to TP c).

2. Then you have to calculate the current moving through the resistor. V = IR, where voltage is now the drop specifically across the resistor. TP a to TP b. There is no way to measure the resistance of the LED to otherwise help figure out the current. Remember that the current moving through one component in series is the same current for all components along that line.

3. Then you calculate the power.

Voltage drop across the LED multiplied by the current in the system equals power used by the LED.

Your numbers should be close to the sample provided. Plug these numbers into the formula for power:

Watts = Volts × Amperes.

```
0.0270 W = 1.77 V × 0.0153 A
```

Oooh! 27 milliwatts. Brilliant.

Exercise: Defining Work, Force, and Power

Complete answers for these exercises can be found at www.books.mcgraw-hill.com/authors/cutcher.

1. Force is the amount of energy exerted. It is measured in what units? _____

2. Roughly 1 N moves _____ grams upward against the force of gravity (specifically it is _____ g).

```
Work = Force × Distance
```

3. Work is the amount of _____ exerted on an object over a _____.

4. If an object is not moved (or the distance equals zero), how much work is done?

5. The unit used to measure work is _____ _____.

   Another name for the same unit is joules. 1 joule = 1 _____.

6. Both are equal to the amount of work done when moving _____ g __m upward.

7. Power is the _____ that work is being done.

8. The common unit for power is _____, abbreviated as ___.

9. Provide an example where two items have different amounts of power output even though the same force is exerted on them.

You probably already know that 1 W = 1,000 mW (milliwatts). The prefix "milli" represents a thousandth (0.001). You've seen it in chemistry (mL) and physics (mm).

Power for electricity is figured using the formula watts = volts × amperes.

10. The LED sample power output is 0.027 watt. That is _____mW.

11. You have seen in the previous exercise how increasing resistance dampens current. The power output is directly proportionate to the current available. If the supply voltage remains constant and the resistance increases, the current decreases.

Don't breadboard the setup in Figure L46-8. You did this same set of resistors early on, and you can copy the values directly from the exercise sheet you did in Lesson 5.

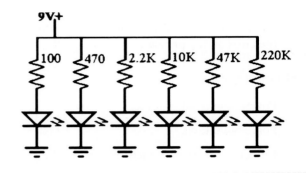

Figure L46-8

Calculate the power available to the LEDs for each setup (Table L46-1).

Table L46-1   Power available

R Value Measured	V+ to Gnd	Current in System $\dfrac{\Omega}{Ohms}$ = Amp	Voltage Drop Across the LED	LED Brightness	Power Used by LED
Copy	Copy	Calculate	$V_{total} - V_R = V_{LED}$	Copy	$W = (V)(I)$
100 Ω	___Volts	0.___ A	___Volts	_____	_0._____W
470 Ω	___Volts	0.___ A	___Volts	Normal	_0._____W
2.2 kΩ	___Volts	0.___ A	___Volts	_____	_0._____W
10 kΩ	___Volts	0.___ A	___Volts	_____	_0._____W
47 kΩ	___Volts	0.___ A	___Volts	_____	_0._____W
220 kΩ	___Volts	0.___ A	___Volts	_____	_0._____W

12. John's car has a 150-watt amp working off 12 volts. His parents have a 22-watt home system that runs off 120 volts. Which system is more powerful?

(Hint: What is the color of George Washington's white horse?) _____

13. What is the current in a 100-watt lightbulb working off 120 volts? _____

People used to say the standard lighting needed to read was the light provided by a 100-watt incandescent light. The same amount of light can be provided by five ultrabright white LEDs at 50 mA each.

14. Compare LEDs for power usage (see Table L46-2).

Table L46-2   Power usage

Type	Voltage	Current	Brightness	Power
Regular red diffuse	1.8 v	30 mA	2,000 mcd	
Regular green diffuse	2.1 v	30 mA	2,000 mcd	
Regular yellow diffuse	2.3 v	30 mA	2,000 mcd	
High-intensity yellow	2.1 v	50 mA	6,500 mcd	
Ultrabright red	2.4v	50 mA	8,000 mcd	

15. Use Ohm's law to calculate the amount of current passing from C to E in Figure L46-9.

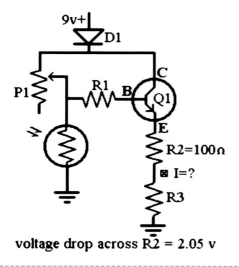

voltage drop across R2 = 2.05 v

Figure L46-9

16. How much power is available at the point between R2 and R3? _____mW

17. Is this enough power to light up an LED? Yes / No

Lesson 47: What Do I Have to Gain? Defining Gain

As you turn the volume control on your radio, the potentiometer's changing resistance isn't just using up some of the power to the speakers. It's actions are much more subtle. Amplifiers use an efficient method to control their power output. Read on and see what you have to gain.

Defining Gain

In an amplifier the input signal is small, and the output signal is big! *Gain* is the ratio that compares input to output.

```
Gain = output/input

Gain = signal_out/signal_in
```

Gain is a statement of a basic ratio. It has no standard unit because the ratio compares identical unit values, such as current.

$$Gain = I_{out}/I_{in}$$

Gain in the NPN Transistor

Set up a modified version of the night light circuit. The schematic is in Figure L47-1 shows a few changes needed to calculate the gain of the NPN 3904 transistor.

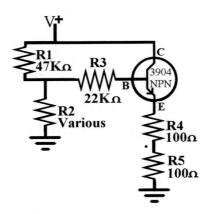

Figure L47-1

1. R1 at 47 kilo-ohms simulates the potentiometer set to midrange.

2. R2 will be various resistors simulating the LDR at different light levels. Bright light would have low resistance.

3. R3 is 22 kilo-ohms. I never did explain before what it did. It restricts current to the base (input) of the transistor.

4. In Figure L47-2, R4 and R5 make a voltage divider. Measuring the voltage drop across R4 makes it an easy task to calculate the current available at the emitter (output).

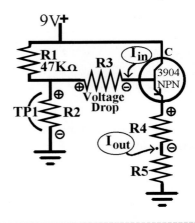

Figure L47-2

How would we calculate gain for this setup?

This is a matter of comparing the current input and current output.

They get set into a ratio, and there is your gain.

$$\text{Gain} = I_{out}/I_{in}$$

Table L47-1 was done with a 12-volt wall adapter as a power supply. The readings here will not necessarily match your readings.

Table L47-1   Ratio and gain

	R2	V TP1	V_{drop} R3	I_{in} @ B V/Ω = A	V_{drop} R4	I_{in} @ R4 V/Ω = A	Gain = I_{out}/I_{in}	Power Used by R4 W = V × A	Power @ E $P_{R4} + P_{R5}$ = P_{total}
1	1 M	7.54 V	2.83	2.83 V/22,000 Ω = 0.129 mA	2.05v	2.05 V/100Ω = 20.5 mA	20.5 mA/ 0.129 mA Gain = 159	2.05 V × 20.5 mA = 0.042 W	0.082 W
2	470 K	7.39 V	2.73v	0.124 mA	2.0	20 mA	161	0.040 W	0.080 W
3	100 K	6.23 V	2.27v	0.105 mA	1.64	16.4 mA	159	26.9 mW	53.2 mW
4	47 K	5.10 V	1.83v	0.083 mA	1.29	12.9 mA	155	16.6 mW	33.2 mW
5	10 K	2.26 V	0.718v	0.033 mA	0.433	4.33 mA	131	1.87 mW	3.74 mW
6	4.7 K	1.26 V	0.299v	0.013 mA	0.152	1.52 mA	116	0.231 mW	0.462 mW

For figuring the power output of the transistor, the following considerations have been made.

1. Since R4 = R5, they both use the same amount of power.

2. The voltage drop across both loads is identical.

3. The current on the same line is unchanged.

4. They use all the power available from the transistor.

5. Therefore $P_{R4} + P_{R5} = P_{Emitter}$

Remember that the LED used 27 mW of power. By the numbers developed in this sample, the LED begins to fade when there is a bit less than 50 kilo-ohms on R2 because from that point on, there is not enough power.

Table L47-2 Readings

NPN 3904	The Readouts Are ± 5%	PNP 3906	
E_{red} to B_{black}	(expect OL)	E_{red} to B_{black}	(expect 0.68)
B_{red} to C_{black}	(expect 0.68	B_{red} to C_{black}	(expect OL)
E_{black} to B_{red}	(expect 0.68)	E_{black} to B_{red}	(expect OL)
B_{black} to C_{red}	(expect OL)	B_{black} to C_{red}	(expect 0.68)

2. Before we move into measuring a complex circuit, let me prove to you that when you calculate the current in one portion of a circuit, that current is the same for the remainder of that circuit. You don't need to make the simple schematic shown in Figure L47-3. All the needed information is there.

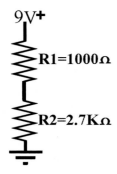

9V+

R1=1000Ω

R2=2.7KΩ

Figure L47-3

First, you need to check the quality of your transistors.

1. Test the quality of your transistors using the DMM diode test. Record your readings in Table L47-2.

   For the circuit-test DMR 2900 multimeter, you first set your DMM to "Continuity" and then press the DC/AC button. The symbol in the top left corner changes from the "beeper" icon to a diode symbol. Many multimeters simply use the continuity tester to give the reading.

Now how much voltage is used across R1 and R2 individually?

Remember the formula to calculate this was

$V_{drop} = V_{total} [R1/(R1 + R2)]$

V_{drop} across R1 = ____

V_{drop} across R2 = ____

3. a. How much current is passing through R1?

   Use Ohm's law: V = IR

   $\dfrac{V}{\Omega} = A$

   V_{drop} across R1/1,000 Ω = current at R1

b. How much current is passing through R2?

V_{drop} across R2/2,700 Ω = current at R2

4. Is the current through R1 and R2 nearly identical? Does IR1 = IR2? Yes/No

5. If you want to set up a test circuit, go ahead and do the real measurements. Remember that you can use three significant figures if you measure the resistance with the DMM directly. Use only two if you don't measure the resistance but just use the color code.

6. Now, using the schematic in Figure L47-2, use the values shown for R2. R2 is the only resistor that changes values. R3 keeps the 22 kΩ value shown in Figure L47-1 (see Table L47-3).

Table L47-3 Readings

	R2	V TP1	V_{drop} R3 $V/\Omega = A$	I_{in} @ B $V/\Omega = A$	V_{drop} R4	I_{in} @ R4 $V/\Omega = A$	Gain = I_{out}/I_{in}	Power Used by R4 $W = V \times A$	Power @ E $P_{R4} + P_{R5}$ = P_{total}
1	1 M								
2	470 K								
3	100 K								
4	47 K								
5	10 K								
6	4.7 K								

Lesson 48: The World Is Analog, So Analog Is the World

Here we finally get to see how an amplifier relates to the real world. But keep in mind that this is not digital. Nothing in the real world is digital. Analog is the world

Amplifiers are analog systems. They deal with changing voltage. Modern technology promotes the use of digital storage of information on CDs and DVDs using 20-bit systems. As well, we have digital transmission of information on cable and the Internet. Good amplifiers feed digital information in, provide digital information out. But amplifiers, by and large, deal with sliding analog voltages to duplicate the sound and light as it occurs naturally.

The Noninverting Amplifier

As shown in Figure L48-1, the NPN transistor works in a very direct manner. As the signal to the base increases, the valve opens. As one goes up, the other goes up in direct proportion. This is a direct relation-

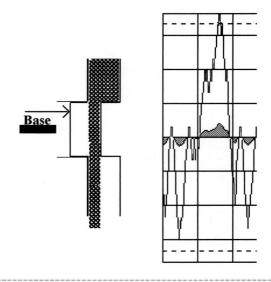

Figure L48-1

ship. The output is proportional with the same voltage as the input. More voltage in creates more voltage out. Voltage is not inverted.

The Inverting Amplifier

Look at Figure L48-2. The PNP transistor has the opposite action of the NPN transistor. As the signal to the base increases, the valve closes. As one goes up, the other goes down in direct proportion. This is an opposing relationship. The output is proportional but opposite voltage as the input.

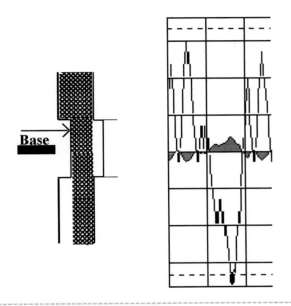

Figure L48-2

The voltage is inverted, But GAIN is the same as the NPN because gain is always stated as an absolute value.

The NPN is a "noninverting" amplifier.

The PNP is an "inverting" amplifier.

Of course, in high tech, we have to use fancy words such as *noninverting*.

Definition: *Inverting*—to turn upside down.

Definition: *Noninverting*—*not* to turn upside down. To leave right side up.

The Op Amp

We are going to be using an 8 pin DIP 741 amplifier on a chip. The basic hookup diagram is shown in Figure L48-3. These are called *Op Amps*, short for operational amplifier. There are many more amplifiers than there are CMOS 4000 series IC chips. Just look at the partial listing at www.books.mcgraw-hill.com/authors/cutcher. Each Op Amp has its own features regarding

- Power input and output
- Response time
- Frequency
- Other factors

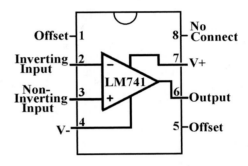

Figure L48-3

The main idea to keep in mind is how any Op Amp works. The Op Amp compares the voltage difference between the two inputs (pins 2 and 3 here) and responds to that difference.

Here we will use the LM741 because of its low power needs and simplicity.

Note that there are both inverting and noninverting inputs.

Exercise: The World Is Analog, So Analog Is the World

1. Which input pin of the LM741 has action similar to the NPN transistor? pin___

2. That action is called what? _____

3. Define the term *inverting amplifier*.

   _____

4. For an inverting amplifier, what is the gain when the output is −100 mV and the input was +10 mV? Be aware that gain is always stated as an absolute value.

   a. −10

   b. +10

   c. −0.1

   d. +0.1

5. The Web site www.books.mcgraw-hill.com/authors/cutcher has a partial list of amplifier ICs that are available on today's market. Would I be right in saying that there is a limited selection of amplifiers available to work with?

   Download a full LM741 data sheet at http://cache.national.com/ds/LM/LM741.pdf. You need to have Adobe PDF reader installed to be able to access this document.

6. On the LM741 data sheet, page 1, the LM741 is manufactured in how many package styles? _____

7. From the LM741 data sheet, page 2, examine the numbers on your chip. Determine which style LM741 you are working with. 741A 741 741C

8. From the LM741 data sheet, page 2, what supply voltage is the LM741 expecting? _____

9. More interesting information from the LM741 data sheet, page 2—your solder pen gets up to 400°C—how long can the LM741 endure direct heat from a solder pen? _____

10. Now turn to page 3 of the LM741 data sheet. The LM741 is considered to be a low-power Op Amp. For example, at rest the typical current draw for this Op Amp is what?

    _____

11. And more info from the LM741 data sheet, page 3—when really pushed, the LM741 consumes a whopping _____mW of power.

12. Turn to page 4 of your LM741 data sheet. How many transistors are packaged into the LM741 IC? _____

Section Fifteen

Exploring the Op Amp

Lesson 49: Alternating Current Compared with Direct Current

Here I explain the necessary analogy comparing alternating current as sound and direct current as wind.

Direct Current

We always hear the terms *direct current* (DC) and *alternating current* (AC). Direct current means that the current in the system is always flowing in one direction. Electronics depends on direct current. Naturally, flowing water has been a good comparison for positive DC voltage V+. Water always flows in one direction—downhill, toward the lowest point. Direct current is easily generated by chemical reactions in batteries. Figure L49-1 shows positive movement of electrons while the scope shows the voltage is above

ground. So the voltage pushes the electrons one way. However, what happens when the voltage is reversed, and pushes the electrons the "other way"? Figure L49-2 shows us how that really works.

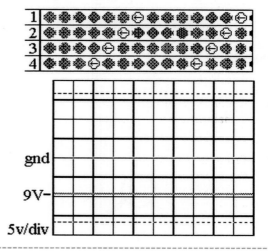

Figure L49-2

Just a note regarding the representations here of DC on a regular oscilloscope. Winscope does not respond to stable DC voltage as a regular oscilloscope would.

With direct current, you can have a positive voltage V+ or a negative voltage V−. Attach a battery backward and you have created V−.

A more effective analogy for DC would be to think of water in a pipe or wind in a tunnel. It can move easily in both directions.

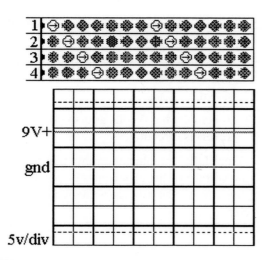

Figure L49-1

191

Alternating Current

Put simply, alternating current is a flow of electrons that keeps alternating, changing directions. The electrons don't get displaced, as they do in DC. Figures L49-3 to L49-5 are frame sets from animations available at www.books.mcgraw-hill.com/authors/cutcher. Those animations show far more clearly the action of what is happening.

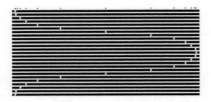

Figure L49-3

Figure L49-4

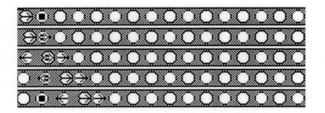

Figure L49-5

But if the electrons don't flow, how is the energy transferred? A great analogy for AC is sound waves. Air particles don't flow as sound travels through the air, but they do move; they vibrate. Sound is a pressure wave that moves through the air. A vibrating string sets one particle bumping into the next.

Simplified, the particles don't get displaced like they do in a wind. They vibrate in place. In fact, when sound is translated into an electronic signal, it is carried as an AC signal as shown in Figure L49-5.

Figure L49-6 represents a regularly oscillating action of electrons in an AC system. This shows a steady AC signal on the scope.

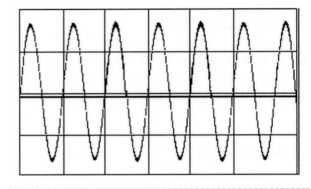

Figure L49-6

AC on the Oscilloscope

Why does a steady AC signal look like this on the oscilloscope screen? The following explains in detail the transformation of real electron movement into a sine wave:

1. This is the actual movement of the electrons, represented in Figure L49-3.

   - Lots of voltage exists as it speeds up in the center. The voltage decreases as it slows down.

   - It actually has 0 volts as it stops and reverses direction.

2. Figure L49-7 graphically shows the increase in the voltage of the electron as it speeds up and slows down. The voltage actually inscribes a circle.

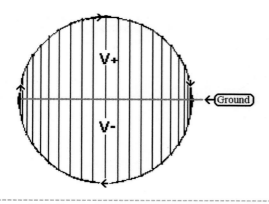

Figure L49-7

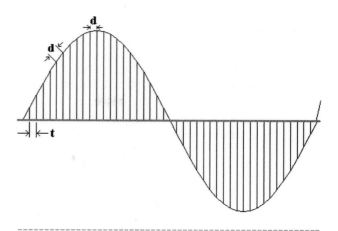

Figure L49-9

Notice two things here.

a. The distance the electron moves along the base line changes as the voltage increases and decreases. Put simply, the bigger the push, the faster it moves. The smaller the push, the slower it moves.

b. Each segment around the outside of the circle actually represents time.

3. And it is not easy to show something that moves backward, because time moves forward. So Figure L49-8 shows the negative voltage movement happening after the positive voltage.

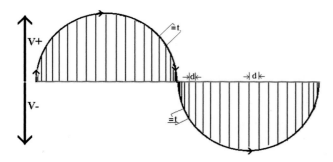

Figure L49-8

4. Time is important. How do we show time? Right now, the time is chasing around on the outside of the circle and distance is marked across the horizontal ground line. But the distance the electrons move is *not* as important to us as timing. So the next thing to do is to impose a time measure onto the horizontal line, replacing distance as the measure. This is shown in Figure L49-9. The amount of voltage

(energy or force) that defined the height remains in place. Obviously, time is marked in evenly spaced increments.

The results? The circular outline changes into the classic sine wave shape.

Note that ground is at the center of the action, the interplay between V+ and V−. As I said, our amplifier deals with sound. You will now learn how to measure with the tools at hand.

Exercise: Exploring the Op Amp

Having a good source of a stable audio signal is vital when you work with audio amplifiers. The workbench tool used for this is a signal generator shown in Figure L49-10. But just like an oscilloscope, you probably don't have one outside of a classroom.

Figure L49-10

However, the computer is a versatile tool that can create and play the test signals that we need as well as show us visually what we need to see.

1. Build the "Test Cord" shown in Figure L49-11. It is similar to building the scope probe but there is no voltage divider. Be sure the black clip is connected to the base of the 1/8" plug's shaft.

Figure L49-11

2. If you don't have a signal generator, you need to use the premade audio signals available at www.books.mcgraw-hill.com/authors/cutcher. These are pure tone wave files. Because wave files are good representations of real analog signals, they can be very large. There are two sets of files. One set is made of three 20-second-long pure tone files. These are for people who need to download using a lower bandwidth. The other set is made of three 2-minute files.

To use these properly, you must create an audio CD and burn these files to the audio CD. Insert the "Test Tone CD" into the computer CD player. Insert the plug into the jack at the front of the CD player to access the signal directly. This output is shown in Figure L49-12.

Then you can take the feed directly from the CD output at the front and feed it to the line input of the sound card. Any media player will work. CDPlayer.exe is very bare-bones and is provided with all Windows operating systems.

3. Disconnect the line from your sound card when you adjust the voltage output. Set your

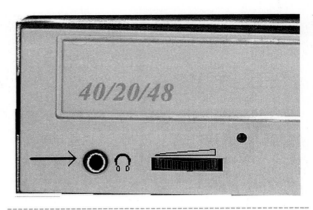

Figure L49-12

DMM to VAC. Adjust the physical volume control on the CD to full volume.

Measure and record the AC voltage shown on the DMM. _____mV. Now adjust the physical volume control to show each setting shown here.

___200 mV   ___100 mV   ___10 mV   ___1 mV

Check mark to verify.

4. Reset the voltage to 200 mV *AC* but set your DMM to V *DC*. No matter how hard you try, you cannot get a measurement of DC voltage from an AC feed. Why doesn't the signal show any DC voltage? You've seen something similar to this before. Look at Figure L49-13.

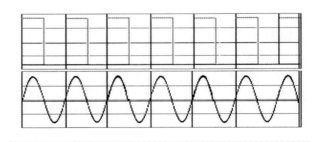

Figure L49-13

Remember when you used the DMM to record the rolldown output from the 4046? The voltage reading at pin 4 was half of the V+ fed to the IC at pin 16. The output is V+ half the time and 0.0 volts for the other half. The signal is switching quickly enough that the DMM averaged the voltage and read it as half of V+.

Now the output is V+ half the time and V−
for the other half. On a DC voltage setting,
the DMM averages the voltage and reads it as
the midpoint between V+ and V−. Hmm?

Could that midpoint possibly be ground? So
the output in DC voltage is zero.

5.   Figure L49-14 shows how to set up Winscope
     to see the tone signals on your computer.

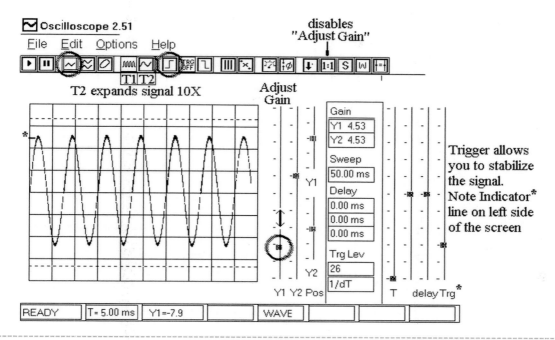

To begin with, a few comments are necessary.

Even though Winscope responds to signals
from inside the computer, you will need to get
used to measuring signals from an outside
source, and feed those into the sound card
microphone or line input.

a.   Attach your "Test Cord" clips to the same
     color clips of the scope probe as shown in
     Figure L49-15.

b.   Insert the scope probe plug into your
     sound card's "microphone input" or "line
     input" jack. Any input greater than 2 volts
     will damage the sound card. The voltage
     divider on the scope probe will ensure that
     does not happen.

c.   Start the signal from the "Test Tone CD."

d.   Physically adjust the small potentiometer
     on the CD to maximum voltage output.
     Software volume controls do not affect the
     volume of the CD's direct output jack.

Figure L49-15

e.   Start Winscope by clicking on the first icon
     button "ON LINE."

f.   Be sure to adjust the settings by clicking
     those shown previously in Figure L49-14.
     Adjust the gain on Y1. When you move the
     "trigger" into place, the signal is stabilized.

6. There are three test tones on the CD. Draw each tone at full volume or 250 mV (whichever is less) on the scope faces provided in Figure L49-16.

7. Adjust the voltage output and redraw each of the tones at 50 mV on the scope faces provided in Figure L49-17.

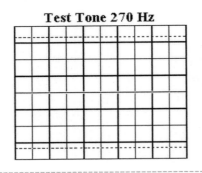

Test Tone 270 Hz

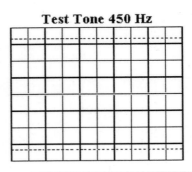

Test Tone 450 Hz

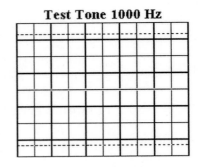

Test Tone 1000 Hz

Figure L49-16

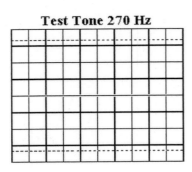

Test Tone 270 Hz

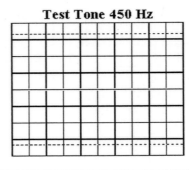

Test Tone 450 Hz

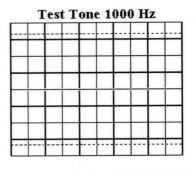

Test Tone 1000 Hz

Figure L49-17

8. Leave the volume adjusted to 50 mV. Place a music CD into the CD player. Feed that signal into your Winscope.

    This is more representative of a normal sound pattern. A pure tone gives a pretty sine wave. Normal sounds and music are not pure tones. They would be very difficult to draw onto an oscilloscope graph.

Lesson 50: AC in a DC Environment

But if alternating current goes forward and backward, how can it be used in a direct current system? How can AC be imposed onto a DC system? Relax.

We can either eat on the ground, or we can pull out a table and use that raised surface. You know enough actually to set up that raised surface in electronics.

The AC Signal in a DC Environment

If AC goes forward and backward, how can we work with AC in a system that uses only V+ from a single 9-volt battery?

To begin, recall the analogy comparing direct current to wind and alternating current to sound. Figure L50-1 shows how sound can be carried in the wind. The animation at www.books.mcgraw-hill.com/authors/cutcher shows the action much more clearly.

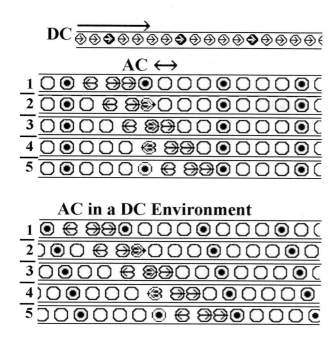

With this in mind, you can carry an AC signal in a DC environment. Here's a larger analogy to explain how. In electricity, you have above ground and below ground. As an analogy, when you go out for a picnic, you can use the ground as a convenient surface, as shown in Figure L50-2.

Figure L50-1

Figure L50-2

Or you can set up a table—in that sense create an artificial surface above the ground as shown in Figure L50-3.

Figure L50-3

Even though the ground is a natural place to put a picnic, an artificial surface like the table is much more convenient. Adjusting the reference for the AC signal in a DC system is much more convenient, too, as you can see in Figure L50-4.

We can create an adjusted reference for an alternating current signal within a DC voltage environment by using a simple voltage divider.

If V+ is 9 volts, then 1/2 of V+ is 4.5 volts.

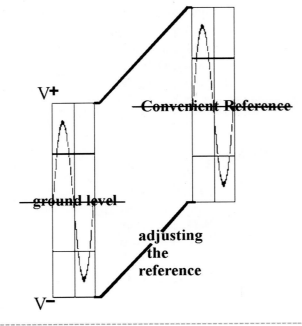

Figure L50-4

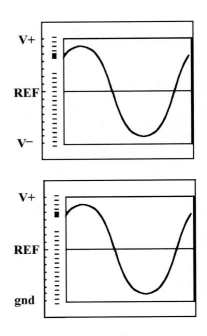

Figure L50-6

In the system shown in Figure L50-5, the adjusted reference acts identically as ground in a 4.5 VAC system.

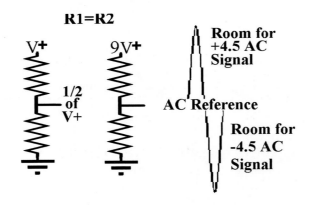

Figure L50-5

How does an AC signal move in a DC environment?

Look closely at the two drawings in Figure L50-6. The image on top shows the AC signal with ground as its natural reference.

The image on the bottom shows the AC signal with an adjusted reference in a DC environment. The current does not affect the signal.

Start Building the Circuit

There needs to be a reference point for carrying the AC signal. Set up the two resistors as shown in Figure L50-7. These will be the voltage dividers used to set your reference point.

1. Take the following measurements:

    a. Total voltage

    b. The reference point voltage

    This allows ____V+ above the reference to act as the positive portion of the AC signal and ____V− below the reference to act as the ____ portion of the AC signal.

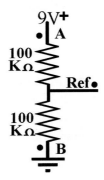

Figure L50-7

Exercise: AC in a DC Environment

1. Define the following terms:

   a. Direct current

   b. Positive voltage V+

   c. Negative voltage V−

2. What is a good analogy for DC that includes both positive and negative DC voltages?

3. Write a good definition of *alternating current*.

4. What is a good analogy for alternating current?

5. What makes alternating current different from direct current?

6. What is the natural reference point for AC?

7. If an AC signal is measured in DC, the DMM displays 0.0 VDC. Explain this.

8. In an AC signal, how much of the signal is above the reference point when compared to the amount of signal below the reference point?

Lesson 51: Comparing Inputs and Audio Coupling

Enough philosophy. Let's set the reference point for the Op Amp and then start building the preamp.

Part 1: When Comparative Inputs Are Equal

Build this circuit displayed in Figure L51-1 on your breadboard (see Table L51-1).

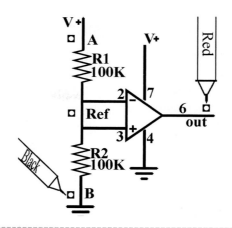

Figure L51-1

Table L51-1   Statements of fact

1. R1 and R2 make the voltage divider that creates reference.	$V_{ref} = \frac{1}{2}$ of V+.
2. The reference voltage V_{ref} is connected to the noninverting input (pin 3).	V_{ref} at pin 3
3. For initial measurement, the voltage provided to the inverting input (pin 2) is identical to the voltage at the noninverting input.	$V_{invert} = V_{noninvert}$
4. The Op Amp output (pin 6) reacts to the voltage difference between the inverting (pin 2) and Noninverting (pin 3) inputs.	
5. Because the difference is ZERO (they are equal), the output of pin 6 is Vref. There is nothing to react to.	$output_{pin\ 6} = V_{ref.}$

Taking Measurements on the Op Amp

1. In this 9-volt DC system, the reference point you are using has _____ V. This allows _____V+ above the reference to act as the positive portion of the AC signal and _____V− below the reference to act as the _____ portion of the AC signal.

2. Is there any difference between the inverting and noninverting input in this setup? Did your measurement support the fourth and fifth "Statements of Fact"? They should have.

Part 2: The Voltage Comparator

Modify the circuit as shown in Figure L51-2. The voltage divider remains attached to the noninverting input, and becomes the reference for the voltage, Vref. Reminder: Disconnect your power when you breadboard.

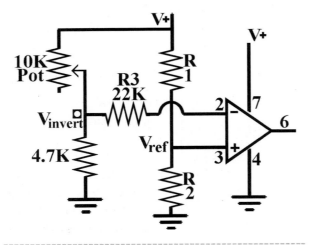

Figure L51-2

The inverting input at pin 2 is connected to the midpoint of a new voltage divider made from a 10-kilo-ohm potentiometer that is balanced against a 4.7-kilo-ohm resistor. Be sure to have the 22 kilo-ohm connected between that midpoint and pin 2.

You have just built a "voltage comparator."

The Op Amp output at pin 6 reacts to any voltage difference between the Inverting input at pin 2 and noninverting input at pin 3. This setup uses the Op Amp to compare two voltages.

It can only determine if

- $V_{inverting}$ is greater than V_{ref}
- $V_{inverting}$ is less than V_{ref}

OK. Let's look closer at it.

3. Use the DMM set to DC voltage. Attach the red probe to pin 3. The black probe goes to ground.

    V_{ref} = _____ V

4. Attach the red probe to V_{invert} and black probe to ground (see Table L51-2).

Table L51-2   Adjust the potentiometer

Adjust the potentiometer so that the midpoint voltage (V_{invert}) is just a bit over V_{ref}.	Adjust the potentiometer so that V_{invert} is even a bit higher.
a. V_{invert} = _____	a. V_{invert} = _____
b. V_{ref} = _____	b. V_{ref} = _____
c. V_{out} = _____	c. V_{out} = _____

5. Table L51-3

Table L51-3   Readings

Adjust the potentiometer so that V_{invert} is just a *under* V_{ref}.	Adjust the potentiometer so that V_{invert} is even a bit lower still.
a. V_{invert} = _____	a. V_{invert} = _____
b. V_{ref} = _____	b. V_{ref} = _____

6. a. When voltage at V_{invert} is greater than V_{ref}, V_{out} at pin 6 becomes _____.

    b. When voltage at V_{invert} is less than V_{ref}, V_{out} at pin 6 becomes _____.

7. Try to adjust the voltage at pin 2 at V_{invert} so that it is the same as V_{ref}.

    If you could do so, what would you expect the V_{out} at pin 6 to be? _____

8. What does a voltage comparator do?

    _____

WAIT! STOP! HOLD IT! Stop and admire the beauty like the Thinker in Figure L51-3. Before we just cruise by this pretty sight, I'd like to point out an item of interest on this sightseeing tour. This one little circuit is the heart of digital inputs.

Figure L51-3

Think about it. You have a reference voltage and an input that could be bigger or smaller. The output responds instantly to the difference. This is your basic NOT gate.

Figure L51-4 shows how a voltage comparator works. An animation of this can be viewed at www.books.mcgraw-hill.com/authors/cutcher.

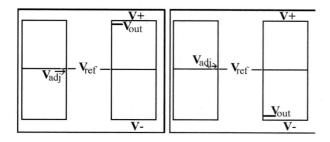

Figure L51-4

So I can hear you mumbling, "OK, so I've got a voltage comparator. But I don't want a voltage comparator. I want an audio amplifier."

Audio Coupling

At this point right now, we will set up and start exploring the preamp. The preamp's purpose is to increase the voltage of the input signal to a usable size.

Modify your Breadboard (Again)

The schematic is displayed in Figure L51-5.

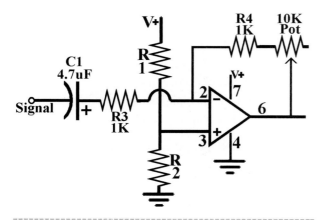

Figure L51-5

a. Remove 4.7-kilo-ohm and 22-kilo-ohm resistors.

b. Add the 4.7-µF cap.

c. Add two 1-K resistors.

d. And rearrange the 10-kilo-ohm pot.

First, recognize that the capacitor is **VITAL** in this position. This is a new way to use a capacitor.

An audio coupler separates the AC signal from DC voltage:

- It isolates the Op Amp from current.
- The AC signal passes through the capacitor.
- The AC signal is automatically set to the Op Amp's reference level.

AC signal passes through the capacitor. This idea is similar to sound passing through a closed window just like it is shown in Figure L51-6. Wind can't pass. Think of how the vibrations of sound pass. The window vibrates, too.

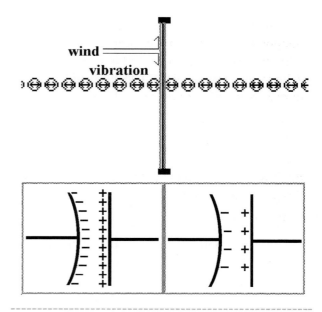

Figure L51-6

The active voltage changes created by the signal on one side of the capacitor affects the charge on the opposite plate of the capacitor.

Notice how the parts have been changed in this setup. There is a feedback loop from the output to the inverting input of pin 2 as shown in Figure L51-7. An animated version of this is available to view at www.books.mcgraw-hill.com/authors/cutcher. The AC signal is already centered around the reference point determined at pin 3.

A portion of the output signal is now being fed back into the system through the potentiometer. This setup allows the potentiometer to act as a volume control. Lesson 52 is dedicated to investigating and explaining the action of the potentiometer and how it acts as the volume control.

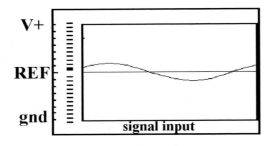
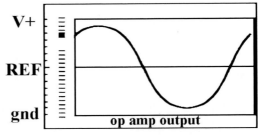

Figure L51-7

Lesson 52: Using Feedback to Control the Gain

I don't want this to sound like a psychology class. Negative feedback? Well, we do need to use negative feedback to control the gain. You just built the Op Amp. Now you get to test it. You will set the gain through its unique feedback system. Actually, it's only a voltage divider applied in a familiar way.

At the end of Lesson 51 you built the preamplifier section. This is where things like volume are controlled. It is so much easier to control the signal when it is small. You are going to build a half-watt intercom system. The potentiometer is rated for one quarter watt. Using the potentiometer to control the volume, the signal to the speakers directly would create enough heat to burn out the pots.

Use Figure L52-1 as your guide as we set up for taking some measurements needed as a foundation for some important concepts.

The red clip of your test probe is set to the signal input. Your black clip is set to ground. You will need to set the input signal to a very small 10 mV AC. Use the DMM VAC to be accurate. The signal is generated by the 1,000 Hz wave file loaded from www.books.mcgraw-hill.com/authors/cutcher. Be

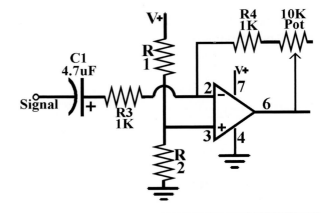

Figure L52-1

sure to use the wave files provided because they are analog signals that have not been digitized or compressed. Use the software volume control to adjust the level of the signal accurately. A mouse wheel is most accurate. Place it on the volume adjust button as shown in Figure L52-2. Wheel it back and forth. You can adjust as little as 2 mV between steps. Use the output for the headphones. Be sure not to use more than 50 mV as your input.

1. Set the pot to minimum resistance.

2. Measure the output signal at pin 6.

3. VAC should be identical to the input and gain = 1.

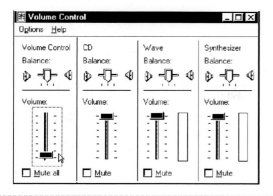

Figure L52-2

4. Set the pot to maximum resistance.

5. Measure the output signal at pin 6.

6. VAC should be 11 times greater than the input (gain = 11).

Recall that gain can be stated as the ratio of voltage to voltage, power to power, and even R1 to R2 in a simple voltage divider.

$$\text{Gain} = \frac{\text{output}}{\text{input}}$$

What to Expect

Ideally, the input signal voltage for this circuit is between 0.010 to 0.015 V *AC*.

- When the pot is set to 0 ohms, the output should equal the input.

- When the potentiometer is set to 0 ohms, then $\frac{R3 + Pot}{R2}$ gives a ratio of 1.

- The voltage divider has a ratio of 1. Gain should be 1.

- When the pot is set to 10,000 ohms, the output should be 11 times greater than the input.

When the potentiometer is set to 10,000 ohms, then $\frac{R3 + Pot}{R2}$ gives a ratio of 11.

The voltage divider has a ratio of 11. Gain should be 11.

How It Works: Feedback to the Inverting Input

Here's a question. "Why does the volume go down when the potentiometer has lower resistance? Wouldn't less resistance mean more current passing? More current passing means more volume?"

The answer is in the fact that the output signal has been inverted and a portion of the inverted signal is now being directed back onto the original signal. It is the same as adding a negative number. Think of it as subtraction. Subtraction is exactly what is being shown in the graphic representation of the signal with feedback in Figure L52-3.

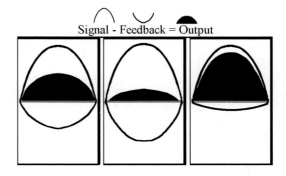

Figure L52-3

The size of the feedback signal increases as the resistance in the potentiometer decreases.

The original signal minus the larger feedback signal results in a smaller output signal.

$$\text{Signal}_{in} - \text{Signal}_{feedback} = \text{Signal}_{output}$$

The graphic in Figure L52-4 more accurately represents a real signal.

Exercise: Using Feedback to Control the Gain

1. Use the setup in Figure L52-5 for collecting the following data (see Table L52-1).

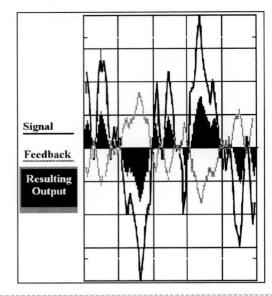

Figure L52-4

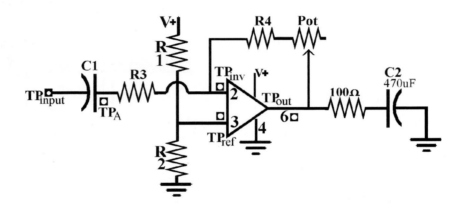

Figure L52-5

Table L52-1   Data

VAC between	Input = 0.0 V_{AC}*	(A) Input = 10.0 mV_{AC} R4 + Pot = 1 kΩ	(B) Input = 10.0 mV_{AC} R4 + Pot = 5 kΩ	(C) Input = 10.0 mV_{AC} R4 + Pot = 10 kΩ
		Measurement Conditions—Voltage and Pot Settings		
TP_{IN} to TP_A				
V_{input} = TP_A to TP_{REF}				
V_{output} = TP_{OUT} to TP_{REF}				
TP_{INVERT} to TP_{REF}				
TP_A to TP_{INVERT}				
Across 100 Ω resistor				

*Input from the headphones output with the volume adjusted to 0.0 volts AC. If the input to the OP Amp is unattached, static in the air will give erroneous and ghostly readings. Recall the unconnected inputs in digital.

2. What was the expected gain using the resistor ratios in the following?

$$Gain = \frac{R4 + Pot}{R3}$$

a. _____

b. _____

c. _____

3. What was the expected gain using the voltage output compared to the voltage input?

$$Gain = \frac{V_{OUT}}{V_{IN}}$$

a. _____

b. _____

c. _____

4. Consider what would happen if R3 were burnt out and had infinite resistance. (>20 MΩ)

a. Write out the expected ratio. _____

b. Think about it. Would that mean the volume would be extra loud, or not at all?

c. Would you be able to control the volume?_____

5. Consider what would happen if R3 were shorted out. In this case, R3 would have 0 ohm, no resistance.

a. State the expected ratio. _____

b. Would you be able to control the volume?

6. Feedback—here you will need to use a bit of math to help explain the real action at the inverting input (pin 2).

Look at the lesson again.

At a gain of 1, you have the least resistance in the pot. How does the "least resistance" provide the quietest volume? Shouldn't less resistance mean more signal?

And that is *exactly* what less resistance means. But it means more signal in the feedback. More signal *subtracted* from the original input.

a. So calculate the amount of current being allowed to "feed back."

Use Ohm's law **V = IR.** You know the voltage and you know the resistance (see Table L52-2).

Table L52-2   Calculations

Current feedback to pin 2
When R4 + Pot = 1 kΩ

Current feedback to pin 2
When R4 + Pot = 5 kΩ

Current feedback to pin 2
When R4 + Pot = 10 kΩ

- -

b. Now calculate the current that is available at the inverting input pin 2. That is a measure of TPA to reference with R4 disconnected so there is no feedback at all. You will get a measure of the full signal. The current at the inverting input pin 2 is

_____.

c. A little subtraction is now in order. For each of the three settings above, calculate the real signal at the inverting input pin 2. Use the simple formula:

full signal − feedback = signal at pin 2

See Table L52-3.

Table L52-3   Calulations

Full signal	Feedback R4 + Pot = 1 kΩ	Signal at pin 2
_____	_____	_____
Full signal	Feedback R4 + Pot = 5 kΩ	
_____	_____	_____
Full signal	Feedback R4 + Pot = 10 kΩ	
_____	_____	_____

- -

7. For this exercise, you need the following pieces from around the house:

- One long and skinny rubber band
- One short rubber band
- One large piece of cardboard
- Three tacks

Now cut the long rubber band in one spot, giving you one long strand.

Cut the short rubber band into two shorter pieces.

Gently stretch and tack the long strand across the cardboard surface. It should be tight enough to pluck.

Now tie a short piece to the center of the long one as demonstrated in Figure L52-6. Secure the end of that tack, too.

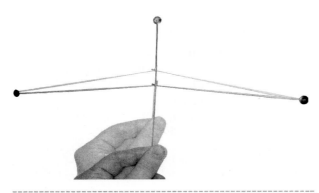

Figure L52-6

The long strand with the small connector altogether represents a signal.

Now attach the second short piece to the center of the long strand. As you tug gently on that loose strand, your input acts like the feedback; your force is being subtracted from the original force on top.

Just like this simple physical demonstration, both the original input and the "inverted" feedback combine to create the signal that is being fed into the Op Amp. The sum of these two combines to create the final output of the Op Amp at pin 6.

Applying the Op Amp; Building the Intercom

Here you'll explore the individual portions of the intercom as you build the system.

Lesson 53: Building a Power Amplifier Controlled by an Op Amp

So the Op Amp is now working as a preamp. The signal voltage has been increased, but there isn't much output at all. Time to crank up the power.

The preamp specifically boosts the voltage to a level that is usable by a power amplifier. Here we will employ transistors as the power amp. The voltage is already there. The power amplifier adds current to the voltage, effectively increasing the power.

Remember: Power = current (I) × voltage (V) or
P = IV

There is a huge variety of amplifier ICs available to achieve the same purpose. But I use transistors because this is a "teaching circuit" that lends itself to explaining concepts common to all amplifiers.

Still, there are many audiophiles that swear by transistors because they provide a different quality output than ICs. But before them, there were the audiophiles that swore by vacuum tubes.

The Power Amplifier

Modify the Circuit

Add these components shown in Figure L53-1 to the preamp you have built. Note the rewiring of the pot and the addition of C2 (1,000 µF) and C3 (470 µF).

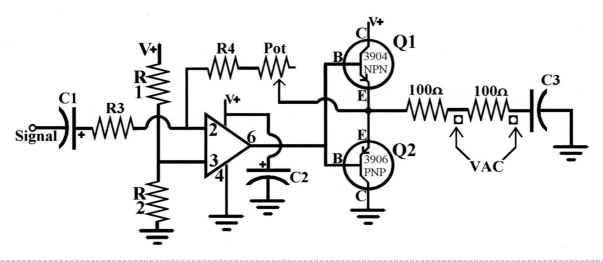

Figure L53-1

The preamp created by the lM741 Op Amp effectively did three things:

1. It amplified the voltage.

2. It limited the amount of current by its very nature.

3. It still provided enough current for the transistors.

The pictures in Figure L53-2 show how the transistors act as power amplifiers. Animated versions of these drawings are available for viewing at www.books.mcgraw-hill.com/authors/cutcher.

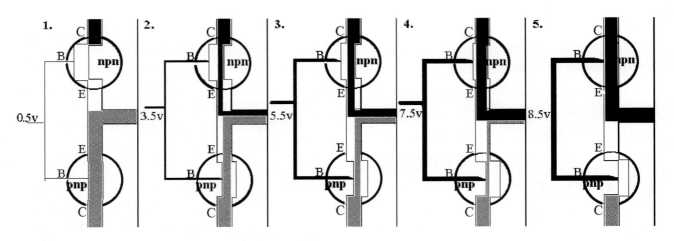

Figure L53-2

The two transistors act in opposition to each other, controlling the signal output. As the valves open and close, they allow for much larger movement of current than allowed previously. The voltage and larger quantities of current create a much more powerful output than the 1M741 could produce by itself.

OK. So I have explained how the power amplifier works. Here is where you will find out how much power your amplifier is really producing. For all of the exercises, use a 1,000 Hz signal input at 10 to 15 mV.

1. Refer back to the measurements you did in Lesson 52. You took all the recordings needed and did all the calculations to get started. Remember:
    Power (watts) = voltage (V) × current (I)

    a. What is the VAC across R3?

    b. What is the current across R3? (V = I × R)

    c. Figure the power across R3 when you had minimum volume (gain = 1). Power at R3 = _____watts

    d. Figure the power across R3 when you had maximum volume (gain = 11).

        Power at R3 = _____watts

2. Now calculate the power output by measuring the VAC available across the 100-ohm resistor. Then measure the VAC across both 100-ohm resistors. The reading should double. What? A voltage divider in AC? With those measurements, you can now calculate the current. How does it feel knowing you are in control of all that power?

    In comparing voltages used, always compare AC voltage to AC voltage. The same goes for comparing DC voltages to each other (see Table L53-1).

Table L53-1 Comparing voltages

	Gain of 1 Minimum Volume	Gain of 11 Maximum Volume
VAC across the 100 Ω load	_____ V	_____ V
V/Ω = A	_____ A	_____ A
Watts = volts × amperes	_____ W	_____ W

3. In Lesson 52, without transistors, what was the power output at maximum volume? _____ W

    In this exercise, with transistors, what was the power output at maximum volume? _____ W

C2—The Capacitor as a Buffer

What is a buffer? A buffer cushions the blow. C2 functions to smooth out the power supply voltage. The voltage demands of the power amplifier create huge fluctuating demands of current. A buffer, or reservoir in the form of a large capacitor, is necessary when using a small power supply such as the 9-volt battery. As the power is quickly tapped, the power fluctuates because it cannot generate current quickly enough. This creates an unstable signal. The capacitor works as a reservoir (the capacitor's basic function). The large capacitor keeps the current and voltage supply steady.

There is an interesting twist here. If you really want to hear this, you can't use a steady tone like we have been using. Keep the same volume you had set for the 1,000-Hz tone. Now insert a music CD. Connect the system to a 9-volt battery. A larger power supply may be able to meet the energy demands of the circuit. Adjust the volume to fairly loud. While you are enjoying your music, pull out C2. The effect is noticeable. Be ready either to pull the power or jab C2 right back into place. The difference is that noticeable.

The Second Audio Coupler

The second audio coupler is in place specifically to isolate the AC output from the DC voltage as shown in Figure L53-3.

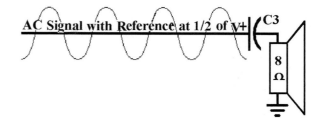

Figure L53-3

In reality, if this capacitor were not in place and the speaker were connected between the output and ground, the entire signal ceases to exist. The AC output signal would be destroyed because the signal and feedback at pin 2 would be referenced to ground in a DC system. They would no longer be floating artificially referenced to a redefined ground at half of the voltage. Take a moment and trace the circuit.

If you want to stop the circuit from working, go ahead and bypass C3 and connect the speaker directly to the connected emitters.

What? It stopped working? Don't say I didn't tell you so.

Lesson 54: Using the Speaker as a Microphone

An intercom has a microphone on one side and a speaker on the other. But you don't have a microphone? You can use the speaker as a microphone. To do this, though, you have to understand a bit about how a speaker works. The speaker was designed as a speaker, not a microphone, as we will use it. It puts out a small signal—tiny.

When we use the speaker as a microphone, the following happens:

- Sound vibrates the cone.

- The cone moves the magnet.

- The magnetic field causes electron movement in the wire.

- This signal is used as the input to the Op Amp.

But still, the speaker is a poor microphone. It wasn't designed to be used as a microphone. But don't take my word for it.

Record this important information.

1. How much of a signal does the speaker produce on your DMM?

    a. Attach the DMM to the speaker. Set it to VAC. Red probe to V+ and black to V−.

    b. Speak into the speaker with your mouth about 2 inches (5 cm) from the cone. It's best to read from printed material rather than sit and repeat "Hellooooo."
    _____mV

    c. The best input signal you can produce is a "pucker" whistle. _____mV

2. How much of a signal does the speaker produce on your oscilloscope?

    a. Hook up your speaker to the test cord, not the one with the voltage divider built in. Plug the test cord directly into the computer sound card microphone input. We are not using the oscilloscope probe. The signal is too small. If we used the voltage divider, it would not register. Use the oscilloscope setup shown here in Figure L54-1.

    b. Speak into the speaker. Make note of the maximum positive and negative voltages on the exercise sheet.

    c. Now whistle into the speaker. You should see a beautiful sine wave. Did you know that your whistle was a pure tone?

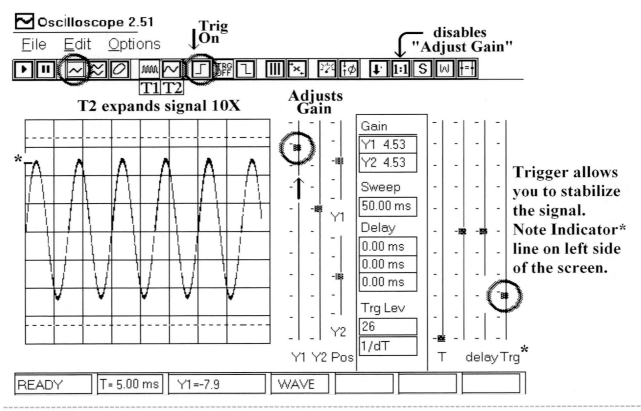

Figure L54-1

d. Draw a representation of your whistle signature onto the scope face provided in Figure L54-2.

Whistle Wave Form

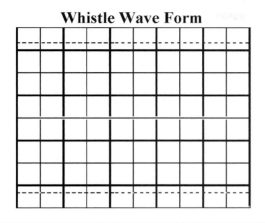

Figure L54-2

3. Applying the speaker in the circuit.

   a. Now set this second speaker into your SBB as shown here in Figure L54-3.

   b. Turn the gain down all the way and use your pucker whistle.

   c. As you increase the volume, the circuit might start screeching at you. Notch the gain backward until the screeching stops. The screeching is caused by feedback demonstrated in Figure L54-4. Sound from the output is reaching the input and multiplying itself.

   d. You can solve this problem in two ways. The first is to give more distance between the output and the input. But right now, you can remove the output speaker. This is an easy solution for the moment. After all, you will need to do some measuring at the output, and don't want the speaker to get in the way.

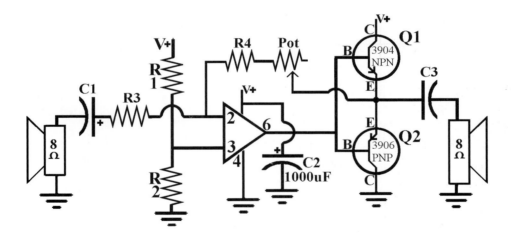

Figure L54-3

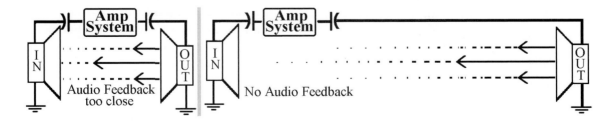

Figure L54-4

4. Scoping the circuit output.

   a. Replace the speaker with the test probe clips, the red clip to C3 (neg), and black to ground. The cord is plugged into the sound card line, or microphone input.

   b. With the volume turned all the way down, whistle and look at the signal on the scope. It should be identical to the signal taken directly from the speaker earlier.

   c. Now turn the volume all the way up. Whistle again. Draw this on you the scope face in Figure L54-5.

Amplified Whistle Wave Form

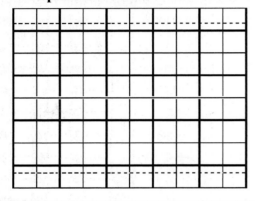

Figure L54-5

5. Check out the AC voltage of the maximum output of your whistling using the DMM. _____ACV Max.

6. Listen to the circuit output.

   a. Now before you put the output speaker back in, solder the ends of the dual wire line to your speaker. That line should be at least 5 feet long. Place it back into the circuit.

   b. Place the input and output speakers as far from each other as possible.

   c. Whistle into the input speaker.

   Hmm. The output is very quiet, even at full volume.

   The last component will take care of that.

Lesson 55: Introducing Transformers and Putting It All Together

Right now the output directly from the speaker is a paltry 1 millivolt. So the output from the amplifier is also very weak. Such a small signal needs to be cleanly amplified even before it is fed to the preamp. To restate, the signal needs to be preamplified before it gets to the preamplifier. Without having a two-stage preamplifier, there just isn't enough signal there to control the power amplifier.

Here I will introduce transformers as a method to preamplify the signal to the preamp. It is frequently done with microphones as well.

Here is a great place to use a transformer. See Figure L55-1.

Figure L55-1   *No no no! Not one of those!*

What we have here is another basic electronic component. Transformers are used in homes in everything from wall adapters to microwaves. Larger versions are vital to the supply and distribution of electric power. You will use a miniature transformer like the one shown in Figure L55-2.

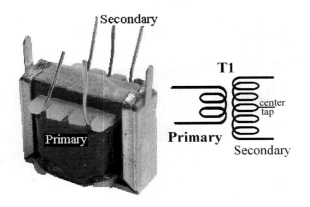

Figure L55-2

How a Transformer Works

Actually, there are two main questions here.

- How does a transformer work?
- What does it transform?

As you saw with the speaker, moving electrons create a magnetic field. What was not mentioned is that the reverse is also true. A moving magnetic field induces (encourages) electrons to move in a conducting wire. The moving electrons in one wire create a magnetic field that induces electrons in nearby wire to move as shown in Figure L55-3.

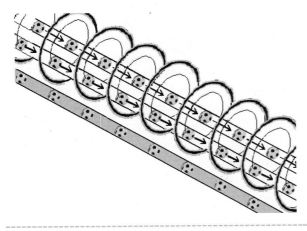

Figure L55-3

Also notice that a second wire is too far away to be influenced by the magnetic field created in the first wire. Animated versions of these figures are available for viewing at www.books.mcgraw-hill.com/authors/cutcher.

Pretty fancy! But better yet, examine Figure L55-4. It depicts how this concept is applied so neatly in electricity and electronics.

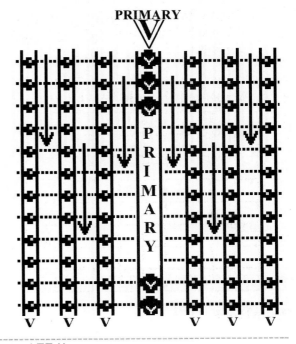

Figure L55-4

If we lay many wires next to the "primary" wire, it will induce (encourage) a voltage in each of the "secondary" wires. But wait. There's more. If each of those secondary wires is connected as one long wire, the voltage induced in each wire is added to the next. This is shown in Figure L55-5.

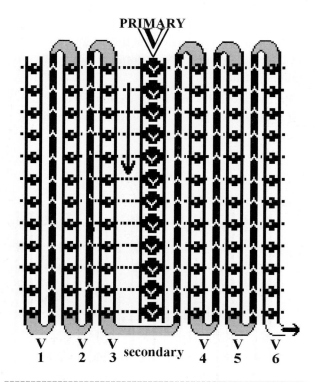

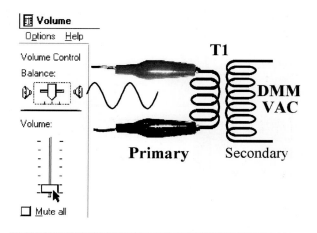

Figure L55-6

Figure L55-6

V
1   V
2   V
3   secondary   V
4   V
5   V
6

Figure L55-5

Checking Out the Audio Transformer

Pull out the miniature audio transformer. It should be wrapped in green. Various colors indicate various uses. Most likely there will be three wires on one side, only two on the other as shown previously in Figure L55-2.

If there are only two wires from each side, continue with the following exercise. It will become obvious which side is primary and which is secondary as you do the work.

The primary side has two wires. The secondary side has three wires. The middle wire on the secondary is referred to as the *center tap*. You can clip and tape it off. We won't use it at all.

Plug the test cord into the headphones output of your sound card. Use the software-based volume control to adjust the 1,000-Hz wave output to 5 mVAC. Now attach the clips to the primary wires of the transformer as shown in Figure L55-6. Measure the output of the two outermost secondary wires (see Table L55-1).

Table L55-1   Output measurement

Input to Primary	Output at Secondary	Ratio of Output VAC / Input VAC
5 mVAC		
10 mVAC		
25 mVAC		
50 mVAC		
100 mVAC		

$$\frac{\text{Output Signal}}{\text{Input Signal}} \quad \text{Average Ratio} = \underline{\hspace{2cm}}$$

In transformers, the fixed gain is referred to as a *step-up* or *step-down* ratio. This ratio actually reflects the physical relationship of the number of primary to secondary windings. A stepup of 8 is created when you have 20 windings in the primary and 160 windings in the secondary. A wall adapter steps the voltage down. For example, one that provides 9 volts from a 120-volt system literally has a ratio of 120 windings in the primary for every 9 in the secondary. That's close to a ratio of 13:1.

The ratio on the audio transformer should be much greater than 1:1. If it is less than 1:1, you have your input connected to the secondary side of the transformer.

Now attach the speaker to the primary side of the audio transformer. Again, use the DMM to set the output at the secondary to measure AC voltage as shown in Figure L55-7.

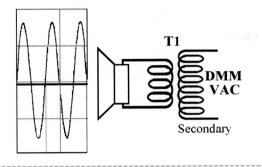

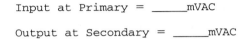

Figure L55-7   *Whistle pucker style for best results.*

```
Input at Primary = _____mVAC

Output at Secondary = _____mVAC
```

Does this match the previous ratios? It should.

Insert the Audio Transformer into the Circuit

1. Insert the audio transformer into the circuit as shown in Figure L55-8.

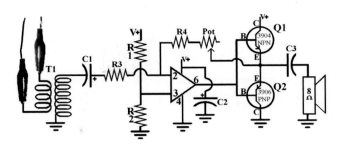

Figure L55-8

2. Now test the entire system using the 1,000-Hz signal. Set the output of your sound card to 5 mVAC by first connecting it to the DMM. Use the software control to set the output signal strength precisely.

3. Measure the AC signal at different points indicated and record the readings in Table L55-2 provided.

Parts list for reference:

R1, R2 100 kΩ   C1 = 4.7 μF

R3, R4 1,000 Ω   C2 = 1000 μF

                       C3 = 470 μF

Table L55-2   AC signal measurements

	5 mVAC Gain = 1	5 mVAC Gain = 11	10 mVAC Gain = 1	10 mVAC Gain = 11
1,000-Hz Sine Wave Tone				
VAC TP input to reference				
VAC TP output to reference				
	5 mVAC Gain = 1	5 mVAC Gain = 11	10 mVAC Gain = 1	10 mVAC Gain = 11
Favorite Music CD				
VAC TP input to reference				
VAC TP output to reference				

Points to keep in mind.

1. This is not a high-fidelity system.

2. Any feed higher than 10 mVAC will probably start sounding lousy because of clipping. The output signal is limited to 4.5 volts above the reference and 4.5 volts below the reference. You can't get more with this system. Try for bigger and you lose most of the signal.

The Intercom System

Remove the test cord and put the speaker to be used as a microphone into place as shown in Figure L55-9.

Now you can use it as an amplifier. Talk into the speaker set up as the microphone. Adjust the gain to get the best quality and volume. At this point, you will find it necessary that the two speakers are separated by at least 5 feet of wire. If they still squeal, turn down the gain or put one speaker on the other side of a sound barrier like a box or door. Take a few more measurements (see Table L55-3).

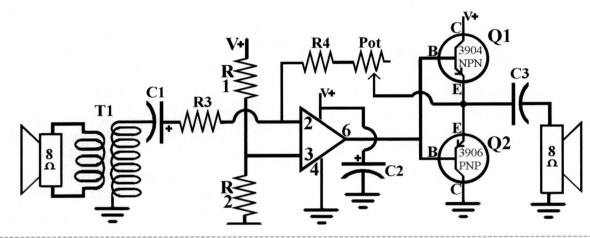

Figure L55-9

Table L55-3   Measurements

Pucker Whistle into Speaker	*mVAC* *Gain = 1*	*mVAC* *Gain = 11*	*mVAC* *Gain = 1*	*mVAC* *Gain = 11*
VAC TP input to reference				
VAC TP output to reference				

Putting It All Together

Lesson 56: Switching to the Two-Way Door Phone

Yes, you now have the first half of the intercom system. But with a flick of a switch, you can reverse the signals, making two halves of an intercom. Then you can finish up.

The Evolution of Switches

Up until now, you have been using momentary contact switches like the ones shown in Figure L56-1.

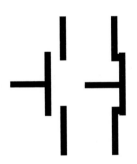

Figure L56-1

Normally open push buttons have contact only when they are pushed and held.

Normally closed push buttons have contact all the time, until they are pushed.

Examples of these can be found in telephones to game controllers.

The single-pole single-throw (SPST) switch displayed in Figure L56-2 has a definite on and off position.

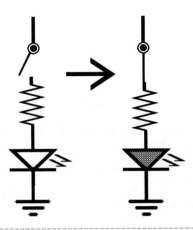

Figure L56-2

It is designed to be used with only one circuit pathway.

The single pole double throw (SPST) switch splits one line into two paths. The example shown in Figure L56-3 shows how a single voltage can be used to power two different items. Sometimes there is an off position set into the middle.

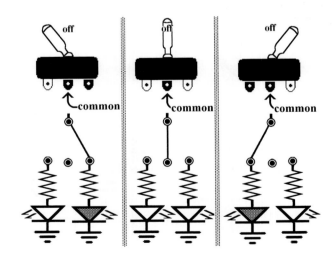

Figure L56-3

217

Notice that the center tab is able to have a connection to either side, depending on the position of the toggle.

The double-pull double-throw (DPDT) switch is like having two SPDT switches glued together side by side, sharing the same toggle. Figure L56-4 displays the commonly found package of the DPDT switch. Note the action of the switch in Figure L56-5. Except for the toggle, the two sides are completely independent of each other.

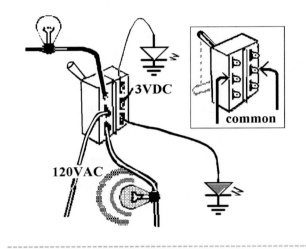

Figure L56-4

The DPDT is found frequently used on stereo systems. They are often paired with indicator lights to show the current function chosen.

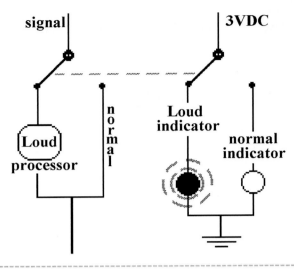

Figure L56-5

But wait . . . aren't stereo systems just that? Stereo!

They have a left *and* a right signal. Each side is completely separate, too! Figure L56-6 shows how two DPDT switches can be made to work together in a big switch. It would be a double, DPDT switch. But let's make life easier. Just call it a 4PDT switch. Figure L56-6 displays the four poles and double throw.

Now that's a switch! They don't come much bigger than that!

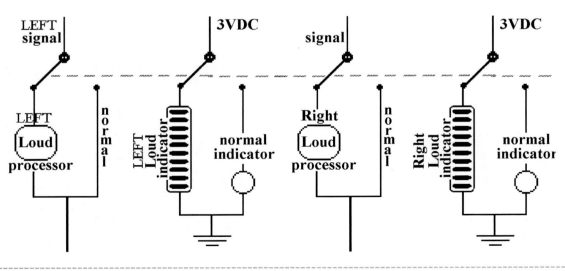

Figure L56-6

Making the Two-way Door Phone

So far, you have the complete system as shown in Figure L56-7.

It uses a speaker as a microphone and you can talk one way. By using the 4PDT switch, you can make it into a two-way system. This is done by rerouting the input and output signals. Figure L56-8 shows the numbering of the tabs for the 4PDT switch. Study the wiring diagram in Figure L56-9 carefully. You don't

want to make a mistake here. Each speaker will have four wires, two on each side.

Recognize that, depending on the position of the switch, only one on each side will be connected at any time. In one position, the speaker on the left acts as the microphone. In the other position, the speaker on the right acts as the microphone.

That is all there is to it!

There is only one switch, and you control it from the inside.

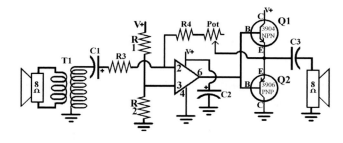

Figure L56-7

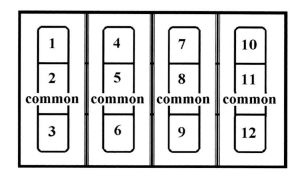

Figure L56-8

Lesson 57: Assembling the Project

See Table L57-1 for project parts list.

Table L57-1

Parts List

R1, R2	100 kΩ
R3, R4	1 kΩ
P1	10 kΩ
C1	4.7 μF
C2	1000 μF
C3	470 μF
T1	Audio transformer (1:50) green
IC1	741 low-power Op Amp
Q1	2N3904 NPN
Q2	2N3906 PNP
Speakers (2)	8 Ω
Switch	4PDT

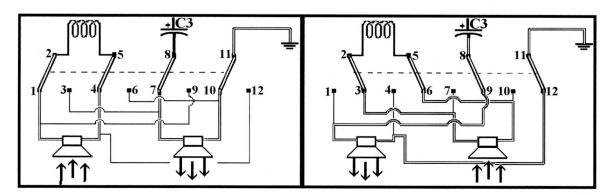

Figure L56-9

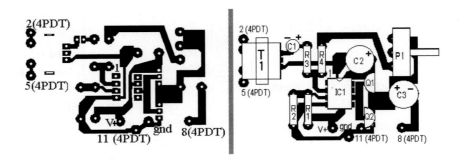

Figure L57-1

Figure L57-1 shows the PCB and parts placement for the door phone.

Assemble the project onto the printed circuit board.

The speaker wiring diagram is shown in Figure L57-2.

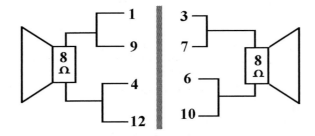

Figure L57-2

Remember to be careful as you do the assembly.

1. Solder the 8-pin chip seat for the 741. Do *not* solder the IC directly.

2. Note the placement of polar components. The only nonpolar components you have in this circuit are resistors.

3. The 3904 and 3906 transistors are not interchangeable in any circuit. Be sure they are properly placed. Overheating the transistors with a solder pen will destroy them. If the circuit doesn't work after soldering, check the transistors. Use the transistor check sheet in the appendix.

4. The potentiometer will not fit directly onto the PCB. Solder short wires to the legs of the potentiometer. Keep them short or they will act like antennas and be the source of a background hummmmmm that shouldn't exist.

One last note. You can easily modify this intercom to provide more power. Purchase the NPN 3904 and PNP 3906 in the larger TO-220 package. Note their pinout may be different from the TO-92 packages. Be sure to attach a large heat sink to the back of each of them. The heat sinks have to be insulated from the rest of the circuit because they have voltage on them. Get a much higher amperage power supply. A wall adapter rated for 500 mA will provide over 4 watts in a 9-volt system. The voltage rating should stay the same. Get 4- or 5-inch 8-ohm speakers rated for 5 watts. Have fun.

Index

ABRA
ELECTRONICS INC.

IN USA: 1320 Route 9, Champlain, N.Y. 12919
Tel.: 800-717-2272 / Fax: 800-898-2272
IN CANADA: 5580 Côte de Liesse, Montreal, Quebec H4P 1A9
Tel.: (514) 731-0117 / Toll Free: 1-800-361-5237 / Fax: (514) 731-0154

Dear Educational Institution or Student,

Congratulations on your purchase of the Evil Genius. As your partner in the study of electronics, we are offering this 10% discount coupon on all additional items purchased from **ABRA ELECTRONICS** when ordering the complete Evil Genius parts kit at the already discounted published price of $55.95 (price US), $65.95 (price CDN). You can also save this coupon to use at a later date if you are only purchasing the parts kit for now. Call, fax or email for your free catalog

CALL FOR YOUR FREE CATALOG!

Who is ABRA?
- *One stop Broad line electronic component distributors specializing in servicing the educational market*
- *#1 supplier of custom made electronic kits across North America. We'll package and label your kits*
- *You simply hand them out to your students*
- *All at no extra charge*

Website: www.abra-electronics.com • Email: kevin@abra-electronics.com